Studies in Fuzziness and Soft Computing

Volume 338

Series editor

Janusz Kacprzyk, Polish Academy of Sciences, Warsaw, Poland
e-mail: kacprzyk@ibspan.waw.pl

About this Series

The series "Studies in Fuzziness and Soft Computing" contains publications on various topics in the area of soft computing, which include fuzzy sets, rough sets, neural networks, evolutionary computation, probabilistic and evidential reasoning, multi-valued logic, and related fields. The publications within "Studies in Fuzziness and Soft Computing" are primarily monographs and edited volumes. They cover significant recent developments in the field, both of a foundational and applicable character. An important feature of the series is its short publication time and world-wide distribution. This permits a rapid and broad dissemination of research results.

More information about this series at http://www.springer.com/series/2941

Antonio Di Nola · Revaz Grigolia
Esko Turunen

Fuzzy Logic of Quasi-Truth: An Algebraic Treatment

Springer

Antonio Di Nola
University of Salerno
Fisciano
Italy

Esko Turunen
Tampere University of Technology
Tampere
Finland

Revaz Grigolia
Ivane Javakhishvili Tbilisi State University
Tbilisi
Georgia

ISSN 1434-9922 ISSN 1860-0808 (electronic)
Studies in Fuzziness and Soft Computing
ISBN 978-3-319-80801-7 ISBN 978-3-319-30406-9 (eBook)
DOI 10.1007/978-3-319-30406-9

Printed on acid-free paper

This Springer imprint is published by Springer Nature
The registered company is Springer International Publishing AG Switzerland

Contents

Chapter 1
Introduction

The question *What is truth?* has intrigued mankind for thousands of years and is related in various senses to our everyday reality and to the disciplines of philosophy, psychology and religion. Asked this question by Pontius Pilate, Jesus Christ responded, *Seek and you shall find.*

In this book, we study truth from the following point of view: we are interested in sentences that are true by some interpretation. The range of interpretation may cover various objects in which interpretation can be realized. In particular, it can be an algebra or a relational system. For classical propositional logic it is the two-element Boolean algebra with an underlying set $\{0, 1\}$, where 1 is interpreted as *true* and 0 as *false*, that is, negation of true. In other words, in classical logic, atomic propositions are evaluated by either 1 or 0. Then any formula of classical logic is evaluated by either 1 or 0, and the value is calculated in the two-element Boolean algebra. Thus there is a one to one relation between classical logic and the two-element Boolean algebra.

Łukasiewicz logic is a non-classical, many-valued logic, originally defined in the early 20th century as a three-valued logic by Jan Łukasiewicz. It was later generalized into n-valued (for all finite n) as well as infinitely-many-valued variants, both propositional and first-order. The infinite-valued version was published in 1930 by Jan Łukasiewicz and Alfred Tarski. This logic belongs to the class known today as t-norm fuzzy logics and substructural logics. Infinite-valued Łukasiewicz logic is a real-valued logic, in which propositions of propositional calculus may be assigned a truth value, not only 0 and 1 but also any real number in between. In other words, we estimate a proposition by some degree of truth that is evaluated by some number between 0 and 1. Łukasiewicz logic takes place in the Hájek framework for mathematical fuzzy logic, because fuzzy logics are based on continuous t-norms, and because Łukasiewicz logic is based on the Łukasiewicz t-norm, although this fact was discovered only several decades after Łukasiewicz' original studies. The algebraic counterpart of infinite-valued Łukasiewicz logic are *MV-algebras*. To give an algebraic proof of the completeness of Łukasiewicz infinite-valued sentential calculus, C.C. Chang introduced *MV*-algebras in 1958 and gave them an equational

A. Di Nola et al., *Fuzzy Logic of Quasi-Truth: An Algebraic Treatment*,
Studies in Fuzziness and Soft Computing 338, DOI 10.1007/978-3-319-30406-9_1

definition. All subvarieties of MV-algebras are known to be finitely axiomatizable and, what is more, each of their axiomatization is also given.

Perfect MV-algebras are an interesting class of *local* MV-algebras. An MV-algebra A is said to be perfect iff for every element a of A, exactly one of a and its complement $\neg a$ is of finite order; that is, in every perfect MV-algebra, for any element a, the meet $a \wedge \neg a$ behaves as an *infinitesimal*. The infinitesimal elements of perfect MV-algebras are *very close* to the falsum, 0. Such elements can be interpreted as models of *quasi falsum*, a negation of *quasi truth*. The first example of a nontrivial perfect MV-algebra, the algebra C, was introduced by C.C. Chang. The algebra C is a notable example of a totally ordered, non-simple MV-algebra. A categorical equivalence is known to exist between MV-algebras and Abelian ℓ-groups with a strong unit. Similarly, there is a categorical equivalence between Abelian ℓ-groups and perfect MV-algebras.

This book aims to study fuzzy-logic-related many-valued logics that are suitable for formalizing the concept of *quasi true*. This suitability is demonstrated by giving a comprehensive account of the basic techniques and results of particular logics and by showing the pivotal role of perfect MV-algebras. These logics are special extensions of Łukasiewicz infinite-valued propositional calculus. In particular, we are interested in truth values that have four gradations. In other words, we have four truth values: true, quasi true, quasi false, and false. Note that if a formula α is quasi true, then $\alpha \odot \alpha$ is also quasi true; and if a formula α is quasi false, then $\alpha \oplus \alpha$ is also quasi false. These truth values have an algebraic origin. The algebras that enable us to introduce such truth values are perfect MV-algebras, that is, MV-algebras that are not semisimple, and whose intersection of maximal ideals (radical of the algebra) is different from $\{0\}$. The non-zero elements of the radical are the infinitesimals. The variety generated by all perfect MV-algebras is generated by a single chain MV-algebra, in fact, the MV-algebra C defined by C.C. Chang.

Perfect MV-algebras are worth exploring for several reasons. To begin with, first order predicate Łukasiewicz logic is known to be incomplete with respect to the canonical set of *truth values* (see [1]); however, it is complete with respect to all *linearly ordered* MV-algebras [2]. Since there are non-simple linearly ordered MV-algebras, we can see that, in this case, the infinitesimal elements of an MV-algebra are allowed to be truth values. In [3], another form of validity is considered for the formulas of first order Łukasiewicz logic. In fact, roughly speaking, a formula α is called *quasi valid* on a model M if for all M-interpretations the value of α is a co-infinitesimal. Therein, it is proved, for a sentence, the equivalence of validity and quasi validity, on all *local* models, that is, on all local MV-algebras. Moreover, the importance of the class of MV-algebras generated by MV-algebras and corresponding to their logic becomes evident when we look at the role infinitesimals play in MV-algebras and in Łukasiewicz logic. Indeed, as said above, pure first order Łukasiewicz predicate logic is not complete with respect to the canonical set of truth values [0, 1]. However a completeness theorem is obtained if the truth values are allowed to vary through all linearly ordered MV-algebras. On the other hand, the incompleteness theorem entails the problem of the algebraic significance of *true but unprovable* formulas. It is significant that the Lindenbaum algebra of first order Łukasiewicz logic is not semisimple, and that valid but unprovable formulas are precisely those whose negations determine the radical of the Lindenbaum algebra, that is, the co-infinitesimals of such algebra. Thus perfect MV-algebras, the variety generated by them, and their logic are intimately related to a crucial phenomenon in first order Łukasiewicz logic.

Secondly, we also stress the fact that in his unpublished note *An MV-algebra for Vagueness* Petr Hájek, in response to some criticism about the logics of vagueness making a sharp break between a true case (value 1) and borderlines cases (value < 1), offered a fuzzy semantics based on a non-standard (non simple) linearly ordered MV-algebra. In fact, for a valuation algebra, Hájek proposed the linearly ordered MV-algebra constructed on the unit interval of the lexicographic product of the real line by itself.

Third, considering the real unit interval $[0, 1]$ as the structure over which to evaluate formulas of a sentential calculus, one has many possibilities. Let us start with Łukasiewicz logic Ł and evaluate formulas by morphisms from the Lindenbaum MV-algebra $\mathfrak{L}$ to $[0, 1]$. We know that Ł is complete with respect to $[0, 1]$. A truth value $x \in [0, 1]$, $x \neq 1$, can be considered as the value of a not-true formula. The distance of x from 1 can be considered to express how close x is to be true. Dually, we can make similar considerations of the falsum 0. Starting from x, assuming $v(\alpha) = x$, where $v(\alpha)$ is the truth value of a formula α, then after a finite number of steps made by the strong (bold) disjunction $\oplus$, such as

$$v(\alpha), v(\alpha \oplus \alpha), v(\alpha \oplus \alpha \oplus \alpha), \ldots$$

we obtain, for every evaluation v,

$$v(\alpha \oplus \cdots \oplus \alpha) = 1$$

and, similarly using the strong (bold) conjunction $\odot$, also

$$v(\alpha \odot \cdots \odot \alpha) = 0.$$

This cannot be a case of Łukasiewicz tautologies α. Indeed, we have $v(\alpha) = 1$ and $v(\alpha \odot \cdots \odot \alpha) = 1$ for all evaluations v. All this is due to the simplicity of $[0, 1]$ and to the semisimplicity of $\mathfrak{L}$. Assume now to evaluate Ł over an ultrapower $^*[0, 1]$ of $[0, 1]$, and assume a formula α such that $v(\alpha)$ is infinitesimally close to 1. We are interested in considering such formulas having a *co-infinitesimal* value for every evaluation. For any of such formulas, their behavior must be intermediate between that of tautologies and that of formulas evaluated into a real number in $[0, 1]$. It is reasonable to consider such formulas as *quasi true*. Therefore, it is an interesting task to explore how to *formalize* such concept of quasi truth and how to develop logics that allow to generalize the concept of truth, that is, to some extent, develop a logic of *approximation*.

Consequently, we are here looking for logics that are extensions of Ł having an evaluation over a non-simple MV-chain. There are several such logics that in different ways concern quasi truth and that, as evaluating algebras, are based on the algebra C or on perfect MV-algebras containing C. In fact, we will focus in this book on several such logics that, roughly speaking, are logics of the concept of quasi true or the concept of *infinitesimally close to the truth.* Notice that the perfect MV-algebra C is a subalgebra of any perfect MV-algebra, which is different from the two-element Boolean algebra; in other words, C is the smallest non-trivial perfect algebra. For example, in the language of the new logic CL (and CL^+ as well), we include a new (constant) connective **c**, which is interpreted as *quasi false*, and hence $\neg\mathbf{c}$ is interpreted as quasi true. Roughly speaking, the constant **c** is a common representative of infinitesimals. Correspondingly, in the signature of the new algebras will appear, beside the MV-algebra operations, a new constant **c**. Thus, in fact, we have infinitely many constants besides 1 and 0: $\mathbf{c}, 2\mathbf{c}, 3\mathbf{c}, \ldots, (\neg\mathbf{c})^3, (\neg\mathbf{c})^2, \neg\mathbf{c}$. As we see, the constant elements form an algebra that is isomorphic to Chang's algebra C. Therefore, the algebra of constant elements should be a subalgebra of all algebras that are models of the logic CL or CL^+.

Another type of logic that is evaluated over the perfect MV-chain C and that we consider in this book is the one recently presented in [2], where such a logic is developed in the context of Pavelka logic. The authors suggest, for example, that logics with infinitesimal truth values are motivated to imitate human reasoning, and they introduce the simplest version of Perfect Pavelka logic, PPL for short. In contrast to Pavelka's [0, 1]-value logic, the logic language of PPL contains only one new truth constant, denoted by the symbol **t** and standing for quasi true. However, unlike the original Pavelka logic on the real unit interval [0, 1], the simplest PPL cannot solve the Sorite Paradox. However, introducing a general perfect MV-algebra-valued Pavelka style logic would solve the problem.

This book requires some acquaintance with classical logic, Łukasiewicz logic, universal algebra, topology, and MV-algebras. However, all the necessary concepts are explained in Chaps. 2, 3, and 4. Chapter 5 deals with local MV-algebras, i.e. with the MV-algebras with exactly one maximal ideal thus containing all infinitesimals. Perfect MV-algebras, a particular class of local MV-algebras, are introduced in Chap. 6, Chap. 7 focuses on the variety generated by perfect MV-algebras, and Chap. 8 examines the representations of perfect MV-algebras. In Chap. 9, we consider the logic L_P, which corresponds to the variety generated by perfect MV-algebras, and in Chap. 10, we introduce a new logic CL by enriching the language of Łukasiewicz logic with a nullary connective interpreted as quasi false. Finally, in Chap. 11, we study Pavelka style fuzzy logic where the set [0, 1] of truth values is replaced by the Chang algebra C.

We are grateful to Mr. Inusah Abdulai, who carefully encoded the LaTeX code of the manuscript according to our instructions. Of course, the authors are responsible for any possible errors.

September 2015 in Tbilisi, Salerno and Tampere

References

1. Scarpellini, B.: Die Nichtaxiomatisierbarkeit des Unendlichwertigen Pradikatenkalkulus von Łukasiewicz. J. Symbolic Logic **27**, 159–170 (1962)
2. Belluce, L.P., Chang, C.C.: A weak completeness theorem for infinite valued predicate logic. J. Symbolic Logic **28**, 43–50 (1963)
3. Belluce, L.P., Di Nola, A.: The *MV*-algebra of first order Łukasiewicz logic. Tatra Mt. Math. Publ. **27**(1–2) 7–22 (2007)

Chapter 2
Basic Notions

2.1 Ordered Sets and Lattices

A binary relation R defined on a set $A \times A$ is a *partial order* on the set A if the following conditions hold for all $a, b, c \in A$:

(i) aRa (reflexivity),
(ii) aRb and bRa imply $a = b$ (antisymmetry),
(iii) aRb and bRc imply aRc (transitivity).
If, in addition, for every $a, b \in A$
(iv) aRb or bRa

then we say R is a *total order on* A. A nonempty set with a partial order on it is called a *Partially ordered set* (*poset* for brevity), and if the relation is a total order then we speak of a *totally ordered set*, or a *linearly ordered set*, or simply a *chain*. When we have a partial order R we use the notation $\leq$ instead of R. In a poset A we use the expression $a < b$ to mean $a \leq b$ but $a \neq b$.

Let A be a subset of a poset P. An element $p \in P$ is an *upper bound* for A if $a \leq p$ for every $a \in A$. An element $p \in P$ is the *least upper bound* of A (l.u.b. of A), or *supremum* of A (sup A) if p is an upper bound of A, and $a \leq b$ for every $a \in A$ implies $p \leq b$ (i.e., p is the smallest among the upper bounds of A). Similarly we can define what it means for p to be a *lower bound* of A, and for p to be the *greatest lower bound of* A (g.l.b. of A), also called the *infimum* of A (inf A). For $a, b \in P$ we say *b covers a*, or a is covered by b, if $a < b$, and whenever $a \leq c \leq b$ it follows that $a = c$ or $c = b$. We use the notation $a \prec b$ to denote a is covered by b. The closed interval $[a, b]$ is defined to be the set of $c \in P$ such that $a \leq c \leq b$, and the open interval (a, b) is the set of $c \in P$ such that $a < c < b$.

A poset L is a *lattice* iff for every $a, b \in L$ both $\sup\{a, b\}$ and $\inf\{a, b\}$ exist (in L).

A. Di Nola et al., *Fuzzy Logic of Quasi-Truth: An Algebraic Treatment*,
Studies in Fuzziness and Soft Computing 338, DOI 10.1007/978-3-319-30406-9_2

2.2 Topological Spaces

A *topological space* is a pair consisting of a set X and some family Ω of subsets of the set X satisfying the following conditions: $\emptyset, X \in \Omega$; if $U_1, U_2 \in \Omega$, then $U_1 \cap U_2 \in \Omega$; if $\Gamma \subset \Omega$, then $\bigcup \Gamma \in \Omega$. The elements of Ω are named *open sets*. The complements of open sets are called *closed sets*. The elements of Ω that simultaneously are open and closed are called *clopen*. In a topological space, we define two operations $\mathcal{I}A$ and $\mathcal{C}A$ as follows:

$$\mathcal{I}A = \bigcup\{B : B \text{ is open subset of } X \text{ and } B \subset A\} \text{ is called } \textit{interior operator}$$

$$\mathcal{C}A = \bigcap\{B : B \text{ is closed subset of } X \text{ and } B \supset A\} \text{ is called } \textit{closure operator.}$$

A class $\mathcal{B}$ of open subsets of X is said to be a *basis* of X if every open subset of X is the union of some sets belonging to $\mathcal{B}$. A class $\mathcal{B}_0$ of open subsets of X is said to be a *subbasis* of X if the class $\mathcal{B}$ composed by the empty set $\emptyset$, the whole space X, and of all finite intersections $B_1 \cap \cdots \cap B_n$ where $B_1, \ldots, B_n \in \mathcal{B}_0$, is a basis of X.

A topological space X is said to be *compact* if, for every indexed set $\{A_t\}_{t\in T}$ of open subsets, the equation $X = \bigcup_{t\in T} A_t$ implies the existence of a finite set $T_0 \subset T$ such that $X = \bigcup_{t\in T_0} A_t$.

A topological space X is said to be T_0-space if, for every pair of distinct points x, y, there exists an open set containing exactly one of them. A topological space X is said to be T_1-space if, for every pair of distinct points x, y, there exist two open sets A and B such that $x \in A$, $y \notin A$ and $x \notin B$, $y \in B$, or equivalently, a topological space X is T_1-space if and only if any finite subset is closed. A topological space X is said to be T_2-space or *Hausdorff space* if, for every pair of distinct points x, y, there exist two disjoint open sets A, B such that $x \in A$ and $y \in B$ and $A \cup B = X$.

2.3 Universal Algebras

The main part of this section is taken from [1].

A *language* (or *type*) of algebras is a set $\mathcal{F}$ of *function symbols* such that a non-negative integer n is assigned to each member f of $\mathcal{F}$. This integer is called the *arity* (or *rank*) of f, and f is said to be an n-ary function symbol. The subset of n-ary function symbols in $\mathcal{F}$ is denoted by $\mathcal{F}_n$.

If $\mathcal{F}$ is a language of algebras then an *algebra* of type $\mathcal{F}$ is an ordered pair (A, F) where A is a nonempty set and F is a family of finitary operations on A indexed by the language $\mathcal{F}$ such that corresponding to each n-ary function symbol f in $\mathcal{F}$ there is an n-ary operation f^A on A. The set A is called the *universe* (or *underlying set*) of (A, F), and the f^A are called the *fundamental operations* of the algebra. We prefer to write just f for f^A and represent an algebra as its underlying set A. If F is finite, say $F = \{f_1, \ldots, f_k\}$ we write $(A, f_1, \ldots, f_k)$. An algebra A is *finite* if $|A|$ is finite, and *trivial* if $|A| = 1$.

Examples:

(1) *Groups*. A *multiplicative group* **G** is an algebra $(G, \cdot, ^{-1}, 1)$ with a binary, a unary, and nullary operations in which the following identities are true:

G1. $x \cdot (y \cdot z) = (x \cdot y) \cdot z,$
G2. $x \cdot 1 = 1 \cdot x = x,$
G3. $x \cdot x^{-1} = x^{-1} \cdot x = 1.$

A group **G** is *Abelian (or commutative)* if the following identity is true:

G4. $x \cdot y = y \cdot x.$

A *additive group* **G** is an algebra $(G, +, -, 0)$ with a binary, a unary, and nullary operations in which the following identities are true:

G'1. $x + (y + z) = (x + y) + z,$
G'2. $x + 0 = 0 + x = x,$
G'3. $x + (-x) = -x + x = 0.$

A group **G** is *Abelian (or commutative)* if the following identity is true:

G'4. $x + y = y + x.$

Groups are generalized to semigroups and monoids.

(2) **Semigroups and Monoids**. A *semigroup* is a groupoid $(G, \cdot)$ in which (G1) is true. It is commutative (or Abelian) if (G4) holds. A *monoid* is an algebra $(M, \cdot, 1)$ with a binary and a nullary operations satisfying (G1) and (G2).

In additive case, we have: a *semigroup* is a groupoid $(G, +)$ in which (G'1) is true. It is commutative (or Abelian) if (G'4) holds. A monoid is an algebra $(M, +, 0)$ with a binary and a nullary operation satisfying (G'1) and (G'2).

(3) **Lattices**. A *lattice* is an algebra $(L, \vee, \wedge)$ with two binary operations which satisfies the following identities:

L1. $x \vee y = y \vee x, \ x \wedge y = y \wedge x,$
L2. $x \vee (y \vee z) = (x \vee y) \vee z, \ x \wedge (y \wedge z) = (x \wedge y) \wedge z,$
L3. $x \vee x = x, \ x \wedge x = x,$
L4. $x = x \vee (x \wedge y), \ x = x \wedge (x \vee y).$

(4) **Bounded Lattices**. An algebra $(L, \vee, \wedge, 0, 1)$ with two binary and two nullary operations is a *bounded lattice* if it satisfies:

BL1. $(L, \vee, \wedge)$ is a lattice
BL2. $x \wedge 0 = 0, \ x \vee 1 = 1.$

(5) **Boolean Algebras**. A *Boolean algebra* is an algebra $(B, \vee, \wedge, \neg, 0, 1)$ with two binary, one unary, and two nullary operations which satisfies:

B1. $(B, \vee, \wedge)$ is a distributive lattice,
B2. $x \wedge 0 = 0, \ x \vee 1 = 1,$
B3. $x \wedge \neg x = 0, \ x \vee \neg x = 1.$

Also, we can define Boolean algebras in another signature. Namely, a *Boolean algebra* is an algebra $(B, \vee, \wedge, \neg)$ with two binary, one unary which satisfies:

B1. $(B, \vee, \wedge)$ is a distributive lattice,
B'2. $x \wedge (y \wedge \neg y) = y \wedge \neg y, \ x \vee (y \vee \neg y) = y \vee \neg y$,
B'3. $y \vee \neg y = x \vee \neg x, \ y \wedge \neg y = x \wedge \neg x$.
In this case we denote $x \vee \neg x$ by 1, and $x \wedge \neg x$ by 0.

(6) **Heyting Algebras**. An algebra $(H, \vee, \wedge, \rightarrow, 0, 1)$ with three binary and two nullary operations is a *Heyting algebra* if it satisfies:

H1. $(H, \vee, \wedge)$ is a distributive lattice,
H2. $x \wedge 0 = 0, \ x \vee 1 = 1$,
H3. $x \rightarrow x = 1$,
H4. $(x \rightarrow y) \wedge y = y, \ x \wedge (x \rightarrow y) = x \wedge y$,
H5. $x \rightarrow (y \wedge z) = (x \rightarrow y) \wedge (x \rightarrow z), \ (x \vee y) \rightarrow z = (x \rightarrow z) \wedge (y \rightarrow z)$.

(7) **Gödel algebras**. An algebra $(G, \vee, \wedge, \rightarrow, 0, 1)$ with three binary and two nullary operations is a *Gödel algebra algebra* if it satisfies:

GH1. $(G, \vee, \wedge, \rightarrow, 0, 1)$ is a Heyting algebra,
GH2. $(x \rightarrow y) \vee (y \rightarrow x) = 1$.

(8) ***BL*-algebras**. An algebra $A = (A, \wedge, \vee, \odot \rightarrow, 0, 1)$ with four binary and two nullary operations is an *BL-algebra* if it satisfies:

BL1. $(A, \wedge, \vee, 0, 1)$ is a lattice with the largest element 1 and the least element 0 (with respect to the lattice ordering $\leq$),
BL2. $(A, \odot, 1)$ is a commutative semigroup with the unit element 1,
BL3. for all $x, y, z \in A$, $x \odot y \leq z$ iff $x \leq y \rightarrow z$,
BL4. for all $x, y \in A$, $x \wedge y = x \odot (x \rightarrow y)$,
BL5. for all $x, y \in A$, $(x \rightarrow y) \vee (y \rightarrow x) = 1$.

(9) ***MV*-algebras**. An algebra $A = (A, 0, \neg, \oplus)$ with one binary and one unary and one nullary operations is an *MV-algebra* if it satisfies:

MV1. $(A, 0, \oplus)$ is an abelian monoid,
MV2. $\neg\neg x = x$,
MV2. $x \oplus \neg 0 = \neg 0$,
MV3. $y \oplus \neg(y \oplus \neg x) = x \oplus \neg(x \oplus y)$.

We set $1 = \neg 0$ and $x \odot y = \neg(\neg x \oplus \neg y)$. We shall write ab for $a \odot b$ and a^n for $\underbrace{a \odot \cdots \odot a}_{n \text{ times}}$, for given $a, b \in A$. Every MV-algebra has an underlying ordered structure defined by

$$x \leq y \text{ iff } \neg x \oplus y = 1.$$

Then $(A; \leq, 0, 1)$ is a bounded distributive lattice.

Moreover, the following property holds in any MV-algebra:

$xy \leq x \wedge y \leq x, y \leq x \vee y \leq x \oplus y.$

(10) **Wajsberg algebras**. An alternative way to define MV-algebras is to start from Wajsberg algebras. An algebra $A = (A, \to, ^*, 1)$ with one binary and one unary and one nullary operation is a *Wajsberg algebra* if it satisfies:

W1. $1 \to x = x$,
W2. $(x \to y) \to [(y \to z) \to (x \to z)] = 1$,
W3. $(x \to y) \to y = (y \to x) \to x$,
W4. $(x^* \to y^*) \to (y \to x) = 1$.

MV-algebras and Wajsberg algebras are in one-to-one correspondence: any MV-algebra satisfies the Wajsberg axioms by stipulations $\neg x = x^*$, $x \oplus y = a^* \to x$, $0 = 1^*$. Also the converse holds; by defining in a Wajsberg algebra $x^* = \neg x$, $x \oplus y = x^* \to y$, $0 = 1^*$ we obtain an MV-algebra.

Let A and B be two algebras of the same type. Then B is a *subalgebra* of A if $B \subset A$ and every fundamental operation of B is the restriction of the corresponding operation of A. Equivalently, a subalgebra of A is a subset B of A which is closed under the fundamental operations of A, i.e., if f is a fundamental n-ary operation of A and $a_1, \ldots, a_n \in B$ we would require $f(a_1, \ldots, a_n) \in B$.

Assume $f : A^n \to A$ is an n-ary operation. The relation $\sim$ is a *congruence* if, for all $a_1, \ldots, a_n, b_1, \ldots, b_n \in A$, $a_i \sim b_i$, $i = 1, \ldots, n$ implies $f(a_1, \ldots, a_n) \sim f(b_1, \ldots, b_n)$.

For A an algebra and $a_1, \ldots, a_n \in A$ let $\theta(a_1, \ldots, a_n)$ denote the congruence generated by $\{(a_i, a_j) : 1 \leq i \leq j\}$, i.e., the smallest congruence such that $a_1, \ldots, a_n$ are in the same equivalence class. The congruence $\theta(a_1, a_2)$ is called a *principal congruence*. For arbitrary $X \subset A$ let $\theta(X)$ be defined to mean the congruence generated by $X \times X$.

Suppose A and B are two algebras of the same type. A mapping $h : A \to B$ is called a *homomorphism* from A to B if

$$h(f(a_1, \ldots, a_n)) = f(h(a_1), \ldots, h(a_n))$$

for each n-ary f in F and each sequence $a_1, \ldots, a_n$ from A. If, in addition, the mapping h is onto then B is said to be a *homomorphic image* of A, and h is called an *epimorphism*. An *isomorphism* is a homomorphism which is one-to-one and onto. In case $A = B$ a homomorphism is also called an *endomorphism* and an isomorphism is referred to as an *automorphism*.

Let A and B be of the same type. A function $h : A \to B$ is an *embedding* of A into B if h is one-to-one homomorphism (such an h is also called a *monomorphism*). For brevity we simply say '$h : A \to B$ is an embedding'. We say A can be embedded in B if there is an embedding of A into B.

Let $f : A \to B$ be a homomorphism. Then the *kernel* of f, $ker(f)$ defined by $ker(f) = \{(a, b) \in A^2 : f(a) = f(b)\}$ is a congruence on A. The set of all congruences of an algebra A forms a lattice $Con A$ with the top element ∇ and the bottom element Δ.

The mapping $\pi_i : A_1 \times A_2 \to A_i$, $i \in \{1, 2\}$, defined by $\pi_i((a_1, a_2)) = a_i$, is called the *projection map on* the ith coordinate of $A_1 \times A_2$.

An algebra A is *(directly) indecomposable* if A is not isomorphic to a direct product of two nontrivial algebras.

We easily generalize the definition of $A_1 \times A_2$ as follows. Let $(A_i)_{i \in I}$ be an indexed family of algebras of the same type. The *(direct) product* $A = \prod_{i \in I} A_i$ is an algebra with universe $\prod_{i \in I} A_i$ and such that for a n-ary fundamental operation f and $a_1, \ldots, a_n \in \prod_{i \in I} A_i$

$$f(a_1, \ldots, a_n)(i) = f(a_1(i), \ldots, a_n(i))$$

for $i \in I$, i.e., f on A is defined coordinate-wise. As before we have *projection maps*

$$\pi_j : \prod_{i \in I} A_i \to A_j$$

for $j \in I$ defined by

$$\pi_j(a) = a(j)$$

which give surjective homomorphisms

$$\pi_j : \prod_{i \in I} A_i \to A_j.$$

An algebra A is a *subdirect product* of an indexed family $(A_i)_{i \in I}$ of algebras if (i) A is a subalgebra of $\prod_{i \in I} A_i$ and (ii) $\pi_i(A) = A_i$ for each $i \in I$. An embedding $h : A \to \prod_{i \in I} A_i$ is *subdirect* if $h(A)$ is a subdirect product of the A_i.

Proposition 2.1 [2] *If $\theta_i \in Con A$ for $i \in I$ and $\bigcap_{i \in I} \theta_i = \Delta$, then the natural homomorphism $h : A \to \prod_{i \in I} A/\theta_i$ defined by $h(a)(i) = a/\theta_i$ is a subdirect embedding.*

An algebra A is *subdirectly irreducible* if for every subdirect embedding $h : A \to \prod_{i \in I} A_i$ there is an $i \in I$ such that $\pi_i \circ h : A \to A_i$ is an isomorphism.

Proposition 2.2 [2] *(i) An algebra A is subdirectly irreducible iff A is trivial or there is a minimum congruence in $Con A - \{\Delta\}$.*

(ii) A subdirectly irreducible algebra is directly indecomposable.

(iii) Every algebra A is isomorphic to a subdirect product of subdirectly irreducible algebras (which are homomorphic images of A).

Suppose $\mathbf{K}$ is a class of algebras, and $A, B \in \mathbf{K}$. The $\mathbf{K}$-*coproduct* of A and B is an algebra $A \sqcup B \in \mathbf{K}$ with algebra homomorphisms $i_A : A \to A \sqcup B$, $i_B : B \to A \sqcup B$, such that $i_A(A) \cup i_B(B) \subset A \sqcup B$ generates $A \sqcup B$, satisfying the following universal property: for every algebra $D \in \mathbf{K}$ with algebra homomorphisms $f : A \to D$

and $g : B \to D$, there exists an algebra homomorphism $h : A \sqcup B \to D$ such that $h \circ i_A = f$ and $h \circ i_B = g$. If we change in the definition of coproduct the requirement that the algebra homomorphisms to be injective, then we have the definition of *free product*. The coproduct $A \sqcup B$ coincides with free product if there is an algebra D such that the algebras A and B can be jointly embedded into D [3].

An algebra A is *simple* if $ConA = \{\Delta, \nabla\}$. A congruence θ on an algebra A is maximal if the interval $[\theta, \nabla]$ of $ConA$ has exactly two elements.

We introduce the following operators mapping classes of algebras to classes of algebras (all of the same type):

$A \in I(\mathbf{K})$ iff A is isomorphic to some member of **K**
$A \in S(\mathbf{K})$ iff A is a subalgebra of some member of **K**
$A \in H(\mathbf{K})$ iff A is a homomorphic image of some member of **K**
$A \in P(\mathbf{K})$ iff A is a direct product of a nonempty family of algebras in **K**
$A \in P_S(\mathbf{K})$ iff A is a subdirect product of a nonempty family of algebras in **K**.

A nonempty class **K** of algebras of the same type is called a *variety* if it is closed under subalgebras, homomorphic images, and direct products. If **K** is a class of algebras of the same type let $\mathcal{V}(\mathbf{K})$ denote the smallest variety containing **K**. We say that $\mathcal{V}(\mathbf{K})$ is *the variety generated by* **K**. If **K** has a single member A we write simply $\mathcal{V}(A)$. A variety **K** is finitely generated if $\mathbf{K} = \mathcal{V}(\mathbf{V})$ for some finite set **V** of finite algebras.

Proposition 2.3 [2] *If* **K** *is a variety, then every member of* **K** *is isomorphic to a subdirect product of subdirectly irreducible members of* **K**.

Let X be a set of (distinct) objects called *variables*. Let $\mathcal{F}$ be a type of algebras. The set $T(X)$ of *terms of type* $\mathcal{F}$ over X is the smallest set such that

(i) $X \cup \mathcal{F}_0 \subseteq T(X)$.
(ii) If $p_1, \ldots, p_n \in T(X)$ and $f \in \mathcal{F}_n$, then $f(p_1, \ldots, p_n) \in T(X)$.

Given a term $p(x_1, \ldots, x_n)$ of type $\mathcal{F}$ over some set X and given an algebra A of type $\mathcal{F}$ we define a mapping $p^A : A^n \to A$ as follows:

(1) if p is a variable x_i, then $p^A(a_1, \ldots, a_n) = a_i$ for $a_1, \ldots, a_n \in A$, i.e., p^A is the ith projection map;
(2) if p is of the form $f(p_1(x_1, \ldots, x_n), \ldots, p_k(x_1, \ldots, x_n))$, where $f \in \mathcal{F}_k$, then $p^A(a_1, \ldots, a_n) = f^A(p^A(a_1, \ldots, a_n), \ldots, p^A(a_1, \ldots, a_n))$.

In particular if $p = f \in \mathcal{F}$, then $p^A = f^A$, where p^A is the term function on A corresponding to the term p. (Often we will drop the superscript A).

Given $\mathcal{F}$ and X, if $T(X) \neq \emptyset$ then the *term algebra* of type $\mathcal{F}$ over X has as its universe the set $T(X)$, and the fundamental operations satisfy

$$f^{T(X)}(p_1, \ldots, p_n) \mapsto f(p_1, \ldots, p_n)$$

for $f \in \mathcal{F}_n$ and $p_i \in T(X)$, $1 \leq i \leq n$. ($T(\emptyset)$ exists iff $\mathcal{F}_0 \neq \emptyset$.)

An *identity* (or *equation*) of type $\mathcal{F}$ over X is an expression of the form

$$p = q$$

where $p, q \in T(X)$. Let $Id(X)$ be the set of identities of type $\mathcal{F}$ over X. An algebra A of type $\mathcal{F}$ *satisfies* an identity

$$p(x_1, \ldots, x_n) = q(x_1, \ldots, x_n)$$

(or the identity is *true in* A, or *holds* in A), abbreviated by

$$A \models p(x_1, \ldots, x_n) = q(x_1, \ldots, x_n),$$

or more briefly

$$A \models p = q,$$

if for every choice of $a_1, \ldots, a_n \in A$ we have

$$p^A(a_1, \ldots, a_n) = q^A(a_1, \ldots, a_n).$$

A class $\mathbf{K}$ of algebras satisfies $p = q$, written

$$\mathbf{K} \models p = g,$$

if each member of $\mathbf{K}$ satisfies $p = g$. If Σ is a set of identities, we say $\mathbf{K}$ satisfies Σ, written

$$\mathbf{K} \models \Sigma,$$

if $\mathbf{K} \models p = g$ for each $p = q \in \Sigma$. Given $\mathbf{K}$ and X let

$$Id_{\mathbf{K}}(X) = \{p = q \in Id(X) : \mathbf{K} \models p = g\}.$$

Let Σ be a set of identities of type $\mathcal{F}$, and define $M(\Sigma)$ to be the class of algebras A satisfying Σ. A class $\mathbf{K}$ of algebras is an *equational class* if there is a set of identities Σ such that $\mathbf{K} = M(\Sigma)$. In this case we say that $\mathbf{K}$ is *defined*, or *axiomatized*, by Σ.

Proposition 2.4 [1] *If* $\mathbf{V}$ *is a variety and* X *is an infinite set of variables, then* $\mathbf{V} = M(Id_{\mathbf{V}}(X))$.

Let $\mathbf{V}$ be a variety. An algebra $A \in \mathbf{V}$ is said to be a *free algebra* over $\mathbf{V}$, if there exists a set $A_0 \subset A$ such that A_0 generates A and every mapping f from A_0 to any algebra $B \in \mathbf{V}$ is extended to a homomorphism h from A to B. In this case A_0 is said to be *the set of free generators* of A. If the set of free generators is finite, then A is said to be a *free algebra of finitely many generators*. We denote a free algebra A with $m \in (\omega + 1)$ free generators by $F_{\mathbf{V}}(m)$. We shall omit the subscript **V** if the variety **V** is known. We can also define the m-generate free algebra A on the generators $g_1, \ldots, g_m$ over the variety **K** in the following way: the algebra A is a free algebra on the generators $g_1, \ldots, g_m$ iff for any m variable identity $p(x_1, \ldots, x_m) = q(x_1, \ldots, x_m)$, the identity holds in the variety **K** iff the equation $p(g_1, \ldots, g_m) = q(g_1, \ldots, g_m)$ is true in the algebra A on the generators [2].

Note that $T(X)$ is indeed generated by X and it is (absolutely) free algebra in the class of all algebras of type $\mathcal{F}$.

Let I be a set. Let $(Su(I), \cup, \cap, ', \emptyset, I)$ be the Boolean algebra of all subsets of I. A subset $F \subset Su(I)$ is said to be *filter* if: (1) $I \in F$, (2) if $X, Y \in F$ then $X \cap Y \in F$, (3) if $X \in F$ and $X \subset Y$ then $Y \in F$. A filter F is *proper* if $F \neq Su(I)$. A proper filter U is called *ultrafilter* if it is a maximal proper filter with respect to the inclusion between filters.

Let $(A_i)_{i \in I}$ be a nonempty indexed family of algebras of type $\mathcal{F}$, and suppose F is a filter over I. Define the binary relation θ_F on $\prod_{i \in I} A_i$ by $(a, b) \in \theta_F$ iff $\{i \in I : a(i) = b(i)\} \in F$.

Proposition 2.5 [1] *For $(A_i)_{i \in I}$ and F as above, the relation θ_F is a congruence on the algebra $\prod_{i \in I} A_i$.*

Given a nonempty indexed family of algebras $(A_i)_{i \in I}$ of type $\mathcal{F}$ and a proper filter F over I, define the *reduced product* $\prod_{i \in I} A_i / F$ as follows. Let its universe $\prod_{i \in I} A_i / F$ be the set $\prod_{i \in I} A_i / \theta_F$, and let a/F denote the element a/θ_F. For f an n-ary function symbol and for $a_1, \ldots, a_n \in \prod_{i \in I} A_i$, let

$$f(a_1/F, \ldots, a_n/F) = f(a_1, \ldots, a_n)/F.$$

If **K** is a nonempty class of algebras of type $\mathcal{F}$, let $P_R(\mathbf{K})$ denote the class of all reduced products $\prod_{i \in I} A_i / F$, where $A_i \in \mathbf{K}$.

A reduced product $\prod_{i \in I} A_i / U$ is called an *ultraproduct* if U is an ultrafilter over I. If all the $A_i = A$, then we write A^I/U and call it an *ultrapower* of A. The *class of all ultraproducts of members of* **K** is denoted $P_U(\mathbf{K})$.

A *quasi-identity* is an identity or a formula of the form $(p_1 = q_1 \& \ldots \& p_n = q_n) \to p = q$. A *quasivariety* is a class of algebras closed under I, S, and P_R, and containing the one-element algebras.

Proposition 2.6 [1] *Let* **K** *be a class of algebras. Then the following are equivalent:*

(a) **K** *can be axiomatized by quasi-identities,*
(b) **K** *is a quasivariety,*
(c) **K** *is closed under I, S, P, and P_U and contains a trivial algebra,*
(d) **K** *is closed under ISP_R and contains a trivial algebra, and*
(e) **K** *is closed under $ISPP_U$ and contains a trivial algebra.*

If **K** is a class of algebras of the same type let $\mathcal{QV}(\mathbf{K})$ denote the smallest quasi variety containing **K**. We say that $\mathcal{QV}(\mathbf{K})$ is *the quasi variety generated by* **K**. If

K has a single member A we write simply $\mathcal{QV}(A)$. A quasi variety **K** is finitely generated if $\mathbf{K} = \mathcal{QV}(\mathbf{V})$ for some finite set **V** of finite algebras.

Proposition 2.7 [4] *Given algebras A_i, $i \in I$, of type $\mathcal{F}$, if U is an ultrafilter over I and Φ is any first-order formula of type $\mathcal{F} \cup \{=\}$, then*

$$\prod_{i \in I} A_i/U \models \Phi(a_1/U, \ldots, a_n/U)$$

iff

$$\{i \in I : A_i \models \Phi(a_1(i), \ldots, a_n(i))\} \in U$$

Let **V** be any variety of algebras. An algebra A is said to be *a retract* of the algebra B, if there are homomorphisms $\varepsilon : A \to B$ and $h : B \to A$ such that $h\varepsilon = Id_A$, where Id_A denotes the identity map over A. An algebra $A \in \mathbf{V}$ is called *projective*, if for any $B, C \in \mathbf{V}$, any onto homomorphism $\gamma : B \to C$ and any homomorphism $\beta : A \to C$, there exists a homomorphism $\alpha : A \to B$ such that $\gamma\alpha = \beta$. Notice that in varieties, projective algebras are characterized as retracts of free algebras.

A subalgebra A of $F_{\mathbf{V}}(m)$ is said to be *projective subalgebra* if there exists an endomorphism $h : F_{\mathbf{V}}(m) \to F_{\mathbf{V}}(m)$ such that $h(F_{\mathbf{V}}(m)) = A$ and $h(x) = x$ for every $x \in A$.

2.4 Categories

A category $\mathcal{C}$ consists of the following data:

(i) A set $Ob(\mathcal{C})$ of objects.
(ii) For every pair of objects $a, b \in Ob(\mathcal{C})$, a set $\mathcal{C}(a, b)$ of *arrows*, or *morphisms*, from a to b.
(iii) For all triples $a, b, c \in Ob(\mathcal{C})$, a *composition map* $\mathcal{C}(a, b) \times \mathcal{C}(b, c) \to \mathcal{C}(a, c)$, $(f, g) \mapsto gf = g \circ f$.
(iv) For each object $a \in Ob(\mathcal{C})$, a morphism $1_a \in \mathcal{C}(a, a)$, called the *identity* of a.

These data are subject to the following axioms:
Associativity: $h(gf) = (hg)f$, for all $(f, g, h) \in \mathcal{C}(a, b) \times \mathcal{C}(b, c) \times \mathcal{C}(c, d)$.
Identity: $f = f1_a = 1_b f$, for all $f \in \mathcal{C}(a, b)$.
Disjointness: $\mathcal{C}(a, b) \cap \mathcal{C}(a', b') = \emptyset$, if $(a, b) \neq (a', b')$ in $Ob(\mathcal{C}) \times Ob(\mathcal{C})$.

We usually write $f : a \to b$ to indicate that a morphism f belongs to $\mathcal{C}(a, b)$. In this case the object a is called the *domain*, or *source*, of f and written $domf$, and b is the *codomain*, or *target*, of f and written $codf$.

The typical example is the category **K** of algebras, whose objects are the algebras from **K** and morphisms are homomorphisms between algebras from **K**. The categories of algebras are examples of *concrete* categories, that is, categories in which objects are sets with additional structure, morphisms are structure-preserving functions, and the composition law is ordinary composition of functions.

A morphism $f : a \to b$ in a category $\mathcal{C}$ is an isomorphism if there exists $g : b \to a$ such that $fg = 1_b$ and $gf = 1_a$. Objects a and b are *isomorphic* if there exists an isomorphism between them, in which case we write $a \cong b$.

A subcategory of a category $\mathcal{C}$ is a subset $\mathcal{D}$ of $\mathcal{C}$ that is closed under composition and formation of domains and codomains. We write $\mathcal{D} \subseteq \mathcal{C}$ to indicate that $\mathcal{D}$ is a subcategory of $\mathcal{D}$.

A morphism m in a category $\mathcal{C}$ is *monic*, or a *monomorphism*, if the equality $mf = mg$ implies that $f = g$, for all morphism $f, g \in \mathcal{C}$. A morphism h is *epi*, or an *epimorphism*, if $fh = gh$ implies that $f = g$, for all $f, g \in \mathcal{C}$. In other words, monics are morphisms that are left cancellable, and epis are morphisms that are right cancellable. In a concrete category, (i) injective $\Rightarrow$ monic; (ii) surjective $\Rightarrow$ epi.

If $\mathcal{C}$ and $\mathcal{D}$ are categories, a *functor* $F : \mathcal{C} \to \mathcal{D}$ consists of functions $Ob(\mathcal{C}) \to Ob(\mathcal{C})$ and $Mor(\mathcal{C}) \to Mor(\mathcal{D})$, also denoted by F, such that (i) $F : \mathcal{C}(a, b) \to \mathcal{D}(F(a), F(b))$, for all $a, b \in \mathcal{C}$; (ii) $F(1_a) = 1_{F(a)}$, for all $a \in \mathcal{C}$; (iii) $F(fg) = F(f)F(g)$, for all composable (f, g).

A functor $F : \mathcal{C} \to \mathcal{D}$ is *faithful* if $F : \mathcal{C}(a, b) \to \mathcal{D}(F(a), F(b))$ is injective for all objects $a, b \in \mathcal{C}$. The functor F is *full* if $F : \mathcal{C}(a, b) \to \mathcal{D}(F(a), F(b))$ is surjective for all $a, b \in \mathcal{C}$. A functor is an *embedding* if it is faithful and is an injective function on objects.

If $\mathcal{C}$ is a category, then the *opposite* category $\mathcal{C}^{op}$ has the same objects and morphisms as $\mathcal{C}$, but with $\mathcal{C}^{op}(a, b) = \mathcal{C}(b, a)$, for all objects a and b, and if $f : a \to b$ and $g : b \to c$ in $\mathcal{C}$, then the composition fg in $\mathcal{C}^{op}$ defined to be the composition gf in $\mathcal{C}$.

We write f^{op} for a morphism $f \to \mathcal{C}(b, a)$ whenever we want to regard f as a morphism in $\mathcal{C}^{op}$, so that $f^{op} : a \to b$ in $\mathcal{C}^{op} \Leftrightarrow f : b \to a$ in $\mathcal{C}$.

A *contravariant* functor from a category $\mathcal{C}$ to a category $\mathcal{D}$ is a functor $\mathcal{C}^{op} \to \mathcal{D}$.

We use the notation $F : \mathcal{C} \to \mathcal{D}$ to denote contravariant, as well as ordinary (covariant) functors. Hence, the statement "$F : \mathcal{C} \to \mathcal{D}$ is a contravariant functor" means that F assigns an object $F(a)$ to each object a in $\mathcal{C}$, and F assigns to each morphism $f : a \to b$ of $\mathcal{C}$ a morphism $F(f) : F(b) \to F(a)$ of $\mathcal{D}$, such that $F(fg) = F(g)F(f)$, for all composable pairs of morphisms (f, g) in $\mathcal{C}$. When we need to emphasize that a particular functor is not contravariant, we will call it a *covariant* functor.

A functor $F : \mathcal{C} \to \mathcal{D}$ yields an *equivalence* of categories $\mathcal{C}$ and $\mathcal{D}$ if and only if it is simultaneously:

(i) full; (ii) faithful; and *essentially surjective (dense)*, i.e. each object d in $\mathcal{D}$ is isomorphic to an object of the form $F(c)$, for c in $\mathcal{C}$.

If a category is equivalent to the opposite (or dual) of another category then one speaks of a *duality* of categories, and says that the two categories are *dually equivalent*.

References

1. Burris, S., Sankappanavar, H.P.: A Course in Universal Algebras. The Millenium Edition (2000)
2. Birkhoff, G.: Lattice Theory. Providence, Rhode Island (1967)
3. Malcev, A.I.: Algebraic Systems. Springer (1973). ISBN 0-387-05792-7
4. Los, J.: Quelques remarques theoremes et problemes sur les classes defmissables d'algebres. In: Skolem, T., et al. (eds.) Mathematical interpretation of formal systems. Studies in Logic and the Foundations of Mathematics, pp. 98–113. Amsterdam (1955)

Chapter 3
Classical Sentential Calculus and Łukasiewicz Sentential Calculus

3.1 Classical Sentential Calculus

Logic was established as a formal discipline by Aristotle (384-322 BCE), who gave it a fundamental place in philosophy.

The classical sentential calculus, classical propositional calculus, or classical propositional logic, as it was, and still often is called, takes its origin from antiquity and are due to Stoic school of philosophy (344-262 BCE). The real development of this calculus began only in the mid-19th century and was initiated by the research done by G. Boole. The classical propositional calculus was first formulated as a formal axiomatic system by G. Frege in 1879 [1].

The assumption underlying the formalization of classical propositional calculus are the following:

We deal only with sentences that can always be evaluated as true or false. Such sentences are called *logical sentences* or *propositions*. Hence the name *propositional logic* or *sentential logic*.

The study of any logic L is begun with its language $\mathcal{L}$. The language $\mathcal{L}$ of classical propositional calculus contains a countable set $\mathbf{Var}(L)$ of *propositional variables* $p_1, p_2, \ldots$, *logical connectives* $\rightarrow, \neg, \vee, \wedge, \leftrightarrow$ (read as 'implies', 'not', 'or', 'and', 'if and only if' respectively). Also, there are left and right brackets. The formulas are defined as follows.

(i) A propositional variable is a formula.

(ii) If α and β are formulas, then so are $(\alpha \rightarrow \beta)$, $(\neg\alpha)$, $(\alpha \vee \beta)$, $(\alpha \wedge \beta)$ and $(\alpha \leftrightarrow \beta)$.

(iii) Any formula is given by the above rules.

We will omit some brackets for simplicity. Denote the set of all formulas by $\mathbf{Form}(L)$.

Now we give some explanation about *semantics* of this logic. A formula of classical propositional logic is more than just a meaningless set of symbols; it can represent a logical combination of facts about the universe. We interpret the propositional

A. Di Nola et al., *Fuzzy Logic of Quasi-Truth: An Algebraic Treatment*,
Studies in Fuzziness and Soft Computing 338, DOI 10.1007/978-3-319-30406-9_3

Table 3.1 Truth tables

α	β	$\alpha \to \beta$	$\neg\alpha$	$\alpha \vee \beta$	$\alpha \wedge \beta$	$\alpha \leftrightarrow \beta$
1	1	1	0	1	1	1
1	0	0	0	1	0	0
0	1	1	1	1	0	0
0	0	1	1	0	0	1

variables as basic statements. If we know whether the basic statements are true or false, we can decide whether any logical compound of them is true or false. In such a way we define semantics of classical propositional logic.

Any formula containing the propositional variables $p_1, \ldots, p_n$ can be used to define a function of n variables, i.e. a function from the set $\{1, 0\}^n$ to $\{1, 0\}$, where 1 is understood as *true* and 0 as *false*. So, we define an evaluation to be a function v from the set of formulas to the set $\{1, 0\}$. The evaluation maps each propositional variable to a truth value, which we take to be the truth value of corresponding basic proposition. We also have to specify how the evaluation behaves as formulas built up. This is done by truth tables as given in Table 3.1. According to the principle of truth functionality the truth-values of compounds of a formula uniquely determine the truth value of a compound of the formula. This achieved by defining the *truth functions* of corresponding logical connectives.

We denote the truth functions by the same symbol as logical connectives. Notice that the set $\{0, 1\}$ with the operations $\vee, \wedge, \neg(= \ ')$, 0, 1 forms the two-element Boolean algebra $(\{0, 1\}, \vee, \wedge, \neg, 0, 1)$. By means of the fundamental operations $\vee, \wedge, \neg$ we can define operations $\to, \leftrightarrow$, 0 and 1 in the following way: $x_1 \to x_2 = \neg x_1 \vee x_2$, $x_1 \leftrightarrow x_2 = (x_1 \to x_2) \wedge (x_1 \to x_2)$, $0 = x_1 \wedge \neg x_1$, $1 = x_1 \vee \neg x_1$. This two-element Boolean algebra is the only simple Boolean algebra and any Boolean algebra is a subdirect product of two-element Boolean algebras [2]. In other words the two-element Boolean algebra generates the variety of all Boolean algebras.

We say that α is a *tautology* if $v(\alpha) = 1$ for all evaluations v, and is a *contradiction* if $v(\alpha) = 0$ for all evaluations v; and that α is a *logical consequence* of a set Σ of formulas if every evaluation v which satisfies $v(\beta) = 1$ for all $\beta \in \Sigma$, also satisfies $v(\alpha) = 1$. It is easy to prove the following

Theorem 3.1 *If the formulas α and $\alpha \to \beta$ are tautologies, then β is a tautology.*

Now we define the formal deduction system Cl for classical propositional logic. For this, we use only the connectives $\neg$ and $\to$, since all the others can be expressed in terms of these. Specifically, if we replace all occurrences of $(\alpha \vee \beta)$ by $((\neg\alpha) \to \beta)$, occurrences of $(\alpha \wedge \beta)$ by $(\neg(\alpha \to (\neg\beta)))$, and occurrences of $(\alpha \leftrightarrow \beta)$ by $(\neg((\alpha \to \beta) \to (\neg(\beta \to \alpha))))$, then the value assigned to the formula by any evaluation is not affected.

There are three 'schemes' of *axioms*, namely for any formulas α, β, γ:

(A1) $(\alpha \to (\beta \to \alpha))$
(A2) $((\alpha \to (\beta \to \gamma)) \to ((\alpha \to \beta) \to (\alpha \to \gamma))$
(A3) $(((\neg\alpha) \to (\neg\beta)) \to (\beta \to \alpha))$

There is only one *inference rule—Modus Ponens*: from α and $(\alpha \to \beta)$, infer β.

A *proof* in Cl is a sequence $\beta_1, \ldots, \beta_k$ of formulas such that for each i, either β_i is an axiom of Cl or β_i is a direct consequence of some of the preceding formulas in the sequence by virtue of the rule of inference of Cl.

A *theorem* of Cl is a formula β of Cl such that β is the last formula of some proof in Cl. Such a proof is called a *proof of β in Cl*.

A formula α is said to be a *consequence* in Cl of a set of Σ of formulas if and only if there is a sequence $\beta_1, \ldots, \beta_k$ of formulas such that α is β_k and, for each i, either β_i is an axiom or β_i is in Σ, or β_i is a direct consequence by the rule Modus Ponens of some of the preceding formulas in the sequence.

When we write out a proof, we precede every formula by the symbol $\vdash$, denoted that it has been proved. If it is a proof from the set Σ, we write $\Sigma \vdash$ on the left.

In the study of a propositional logic L, the following construction is often important: take the set of all formulas in the language $\mathcal{L}$ of L, and partition this set into classes of L-equivalent formulas. In many cases, the set of L-equivalence classes has a natural algebraic structure, which is called the *Lindenbaum algebra for the logic L*. The set $\mathbf{Form}(L)$ of all formulas of a language $\mathcal{L}$ is an universal algebra

$$(\mathbf{Form}(L), \vee, \wedge, \to, \neg)$$

with three binary operations $\vee, \wedge, \to$ and one unary operation $\neg$ —defined as follows: the formulas $(\alpha \vee \beta)$, $(\alpha \wedge \beta)$, $(\alpha \to \beta)$, $(\neg\alpha)$ are the results of the operations $\vee, \wedge, \to$ applied on the formulas α, β, respectively. This algebra is called the *algebra of formulas of the language $\mathcal{L}$*.

The algebra $(\mathbf{Form}(L), \vee, \wedge, \to, \neg)$ is free in the class of algebras

$$(A, f_1, f_2, f_3, f_4)$$

with three binary operations f_1, f_2, f_3 and one unary operation f_4 with all propositional variables $\mathbf{Var}(L)$ being the set of free generators [3].

By an *evaluation* of L in the algebra $(A, \vee, \wedge, \to, \neg)$ we shall understand a mapping $v : \mathbf{Var}(L) \to A$.

Every formula α of k propositional variables in L uniquely determines an operatrion α_A in A, namely a mapping (algebraic polynomial) $\alpha_A : A^k \to A$. To obtain α_A it suffices to interpret the signs $\vee, \wedge, \to, \neg$ in α as signs of the corresponding operations in A, and the propositional variables $p_1, \ldots, p_k$ in α respectively as variables $x_1, \ldots, x_k$ ranging over in A.

Define an equivalence relation $\equiv$ on $\mathbf{Form}(L)$ as follows: $\alpha \equiv \beta$ iff $\vdash \alpha \leftrightarrow \beta$. This equivalence relation is a congruence relation on the algebra of formulas $(\mathbf{Form}(L), \vee, \wedge, \to, \neg)$. Then the factor algebra

$$(\mathbf{Form}(L)/\equiv, \vee, \wedge, \to, \neg)$$

is the Lindenbaum algebra for the logic L, where $\alpha/\equiv \vee\ \beta/\equiv = (\alpha \vee \beta)/\equiv$, $\alpha/\equiv \wedge\ \beta/\equiv = (\alpha \wedge \beta)/\equiv$, $\alpha/\equiv \to \beta/\equiv = (\alpha \to \beta)/\equiv$, $\neg(\alpha/\equiv) = (\neg\alpha)/\equiv$.

The Lindenbaum algebra for classical propositional logic Cl is a Boolean algebra [3], and it is the free Boolean algebra with free generators $x_1, x_2, \ldots$ where $x_1 = p_1/\equiv$, $x_2 = p_2/\equiv, \ldots$. The Lindenbaum algebra for classical propositional logic Cl on n variables $p_1, \ldots, p_n$ is a Boolean algebra, and it is the free Boolean algebra on n generators $x_1, \ldots, x_n$. Notice that in the Lindenbaum algebra the element $(\alpha \to \alpha)/\equiv$ is the top element, which we denote by 1, and the element $(\neg(\alpha \to \alpha))/\equiv$ is the bottom element, which we denote by 0.

The main statements of logics are deduction theorem, soundness and completeness.

Theorem 3.2 (Deduction Theorem) [4] *If Σ is a set of formulas and α and β are formulas, and $\Sigma \cup \{\alpha\} \vdash \beta$, then $\Sigma \vdash \alpha \to \beta$.*

Theorem 3.3 (Soundness) *If α is a theorem of Cl, then α is a tautology.*

Proof The proof immediately follows from the Theorem 3.1 and the fact that every axiom is a tautology. □

Theorem 3.4 (Completeness) *If α is a tautology, then α is a theorem of Cl.*

Proof We give algebraic proof of this assertion. Let us suppose that α is not a theorem of Cl. Then $\alpha/\equiv \neq 1$ in the Lindenbaum algebra $(\mathbf{Form}(L)/\equiv, \vee, \wedge, \to, \neg)$ which is a Boolean algebra. As we know any algebra, and the Lindenbaum algebra $(\mathbf{Form}(L)/\equiv, \vee, \wedge, \to, \neg)$ in particular, is a subdirect product of subdirectly irreducible algebras. But the only subdirectly irreducible Boolean algebra is two-element Boolean algebra. So, there exists a subdirect embedding $h : \mathbf{Form}(L)/\equiv \to \prod_{i \in I} B_i$ such that $h(\mathbf{Form}(L)/\equiv)$ is a subdirect product of an indexed family $(B_i)_{i \in I}$ where all B_i, $i \in I$, are isomorphic to two-element Boolean algebras. So, there exists an element $a \in \prod_{i \in I} B_i$ such that $h(\alpha/\equiv) = a \neq 1$. It means that there exists a projection map $\pi_j : \prod_{i \in I} B_i \to B_j$ for $j \in I$ such that $\pi_j(a) = a(j)$ and $a(j) \neq 1$. Therefore, we have an evaluation $v = \pi_j h g : \mathbf{Form}(L) \to B_j$, where g is the natural homomorphism from $\mathbf{Form}(L)$ onto $\mathbf{Form}(L)/\equiv$ such that $v(\alpha) \neq 1$. From here we conclude that α is not tautology. □

3.2 Łukasiewicz Sentential Calculus

Łukasiewicz logic was originally defined in the early 20th-century by Jan Łukasiewicz as a three-valued logic [5]. It was later generalized to n-valued (for all finite n) as well as infinitely-many valued ($\aleph_0$-valued) variants, both propositional and first-order [6].

The $\aleph_0$-valued version was published in 1930 by Jan Łukasiewicz and Alfred Tarski [7]. It belongs to the classes of t-norm fuzzy logics [8] and substructural logics [9].

The propositional connectives of Łukasiewicz logic are implication $\rightarrow$, negation $\neg$, equivalence $\leftrightarrow$, weak conjunction $\wedge$, strong conjunction $\odot$, weak disjunction $\vee$, strong disjunction $\oplus$.

For Łukasiewicz propositional logic Ł we deal with sentences that can be evaluated with some truth value being in closed interval [0, 1], or, roughly speaking, between true and false. As in classical propositional logic the language $\mathcal{L}$ of Łukasiewicz propositional calculus contains a countable set **Var**(Ł) of *propositional variables* $p_1, p_2, \ldots$, *logical connectives* implication $\rightarrow$, negation $\neg$, equivalence $\leftrightarrow$, weak conjunction $\wedge$, strong conjunction $\odot$, weak disjunction $\vee$, strong disjunction $\oplus$. Also, there are left and right brackets. The formulas are defined as follows.

(i) A propositional variable is a formula.

(ii) If α and β are formulas, then so are $(\alpha \rightarrow \beta)$, $(\neg\alpha)$, $(\alpha \vee \beta)$, $(\alpha \oplus \beta)$, $(\alpha \wedge \beta)$, $(\alpha \odot \beta)$ and $(\alpha \leftrightarrow \beta)$.

(iii) Any formula is given by the above rules.

We will omit some brackets for simplicity. Denote the set of all formulas by **Form**(Ł).

Now we give some explanation about *semantics* of this logic. A formula of Łukasiewicz propositional logic is evaluated by some element of [0, 1]. We interpret the propositional variables as basic statements. If we know whether the basic statements are evaluated by some elements of [0, 1], we can decide whether any logical compound of them is an element of [0, 1]. In such a way we define semantics of Łukasiewicz propositional logic.

Any formula containing the propositional variables $p_1, \ldots, p_n$ can be used to define a function of n variables, i.e. a function from the set $[0, 1]^n$ to [0, 1], where 1 is understood as *absolutely true* and 0 as *absolutely false*. So, we define an evaluation function v from the set of formulas to the set [0, 1]. The evaluation maps each propositional variable to a truth value, which we take to be the truth value of corresponding basic proposition. We also have to specify how the evaluation behaves as formulas built up. According to the principle of truth functionality the truth-values of compounds of a formula uniquely determine the truth value of a compound of the formula. This is achieved by defining the *truth functions* of corresponding logical connectives.

The truth functions corresponding to logical connectives are defined as follows:

Implication: $x \rightarrow y = min\{1, 1 - x + y\}$
Equivalence: $x \leftrightarrow y = 1 - |x - y|$
Negation: $\neg x = 1 - x$
Weak Conjunction: $x \wedge y = min\{x, y\}$
Weak Disjunction: $x \vee y = max\{x, y\}$
Strong Conjunction: $x \odot y = max\{0, x + y - 1\}$
Strong Disjunction: $x \oplus y = min\{1, x + y\}$.

We say that Łukasiewicz formula $\alpha \in \mathbf{Form}(Ł)$ is a *tautology* if $v(\alpha) = 1$ for all evaluations $v : \mathbf{Form}(Ł) \to [0, 1]$, and is a *contradiction* if $v(\alpha) = 0$ for all evaluations v; and that α is a *logical consequence* of a set Σ of formulas if every evaluation v which satisfies $v(\beta) = 1$ for all $\beta \in \Sigma$, also satisfies $v(\alpha) = 1$. It easy to prove the following

Theorem 3.5 *If the formulas* α *and* $\alpha \to \beta$ *are tautologies, then* β *is a tautology.*

The truth function $\odot$ of strong conjunction is the Łukasiewicz t-norm and the truth function $\oplus$ of strong disjunction is its dual t-conorm. The truth function $\to$ is the residuum of the Łukasiewicz t-norm. Since all these truth functions are operations on $[0, 1]$ we can convert the set $[0, 1]$ with operations $\odot, \oplus, \neg, 0, 1$ into the algebra $([0, 1], \odot, \oplus, \neg, 0, 1)$ that is called standard MV-algebra for Łukasiewicz logic. The type of this algebra is $(2, 2, 1, 0, 0)$ like in Boolean algebras. In this algebra we have the following identities:

(1) $x \to y = \neg x \oplus y,\ x \oplus y = \neg x \to y,$
(2) $\neg\neg x = x,$
(3) $\neg(x \oplus y) = \neg x \odot \neg y,$
(4) $\neg(x \odot y) = \neg x \oplus \neg y,$
(5) $x \leq x \oplus y,$
(6) $x \oplus y = y \oplus x,$
(7) $x \oplus (y \oplus z) = (x \oplus y) \oplus z,$
(8) $x \wedge y = (x \oplus (\neg y)) \odot y,$
(9) $x \vee y = (x \odot (\neg y)) \oplus y,$
(10) $x \oplus (\neg x) = 1,\ x \odot (\neg x) = 0.$

The original system of axioms for propositional infinite-valued Łukasiewicz logic used implication and negation as the primitive connectives as for classical logic:

$Ł_1$. $(\alpha \to (\beta \to \alpha))$
$Ł_2$. $(\alpha \to \beta) \to ((\beta \to \gamma) \to (\alpha \to \gamma))$
$Ł_3$. $((\alpha \to \beta) \to \beta) \to ((\beta \to \alpha) \to \alpha)$
$Ł_4$. $(\neg\beta \to \neg\alpha) \to (\alpha \to \beta)$.

There is only one *inference rule—Modus Ponens*: from α and $(\alpha \to \beta)$, infer β.

For Łukasiewicz logic we have no deduction theorem in the form that we have in classical logic Cl since in Ł the formula $\alpha \leftrightarrow \alpha^n$ is not a theorem of Ł, where $\alpha^n = \alpha \odot \ldots \odot \alpha$ (n times).

Theorem 3.6 (Deduction Theorem) [8] *For any* $\alpha, \beta \in \mathbf{Form}(L)$ *and a set of formulas* Σ, $\Sigma \cup \{\alpha\} \vdash_Ł \beta$ *iff there exists a positive integer* n *such that* $\Sigma \vdash_Ł \alpha^n \to \beta$.

Theorem 3.7 (Soundness and completeness) [8, 10] *A Łukasiewicz formula* α *is a theorem of Łukasiewicz logic iff* α *is a tautology.*

References

1. Frege, G.: *Begriffsschrift, eine der arithmetischen nachgebildete Formelsprache des reinen Denkens*. Halle: Nebert, L. (1879). Translated as Begriffsschrift, a Formula Language, Modeled upon that of Arithmetic, for Pure Thought. InFrom Frege to Gdel, edited by Jean van Heijenoort. Harvard University Press, Cambridge (1967)
2. Birkhoff, G.: *Lattice Theory*. Providence, Rhode Island (1967)
3. Rasiowa, H., Sikorski, R.: The Mathematics of Metamathematics. PWN-Polish Scientific Publishers, Warszawa (1970)
4. Herbrand, J.: *Recherches sur Ia Theorie de Ia Demonstration*, Travaux de Ia Societe des Sciences et des Letf1es de Varsovie, III, **33**, 33–160 (1930). Logical Writings. Harvard University Press and Reidel (1971)
5. Łukasiewicz, J.: O Logice trojwartociowej, Ruch filozoficzny **5**, 170–171 (1920) (in Polish). English translation: *On Three-Valued Logic*. In: Borkowski, L. (ed.) Selected Works by Jan Łukasiewicz, North Holland, Amsterdam, pp. 87–88 (1970)
6. Hay, L.S.: Axiomatization of the infinite-valued predicate calculus. J. Symbol. Logic **28**, 77–86 (1963)
7. Łukasiewicz, J., Tarski, A.: Untersuchungen über den Aussagenkalkül, Comp. Rend. Soc. Sci. et Lettres Varsovie Cl. III **23**, 30–50 (1930)
8. Hájek, P.: Metamathematics of Fuzzy Logic. Kluwer, Dordrecht (1998)
9. Ono, H.: *Substructural Logics and Residuated Lattices an Introduction*. In: Hendricks, F.V, Malinowski, J. (eds.) Trends in Logic, 50 Years of Studia Logica, Trends in Logic 20, pp. 177–212 (2003)
10. Chang, C.C.: Algebraic analysis of many-valued logics. Trans. Am. Math. Soc. **88**, 467–490 (1958)

Chapter 4
MV-Algebras: Generalities

4.1 *MV*-Algebras

C.C. Chang introduced MV-algebras as algebraic models for Łukasiewicz logic to give its algebraic analysis [1] and proved completeness of Łukasiewicz logic with respect to the variety of all MV-algebras. We give the definition of MV-algebra given originally by C.C. Chang in [1]. An MV-algebra is a system $(A, \oplus, \odot, \neg, 0, 1)$ where A is a nonempty set of elements, 0 and 1 are distinct constant elements of A, $\oplus$ and $\odot$ are binary operations on elements of A, and $\neg$ is a unary operation on elements of A obeying the following axioms.

Ax. 1. $x \oplus y = y \oplus x$. Ax. 1'. $x \odot y = y \odot x$
Ax. 2. $x \oplus (y \oplus z) = (x \oplus y) \oplus z$. Ax. 2'. $x \odot (y \odot z) = (x \odot y) \odot z$.
Ax. 3. $x \oplus \neg x = 1$. Ax. 3'. $x \odot \neg x = 0$.
Ax. 4. $x \oplus 1 = 1$. Ax. 4'. $x \odot 0 = 0$.
Ax. 5. $x \oplus 0 = x$. Ax. 5'. $x \odot 1 = x$.
Ax. 6. $\neg(x \oplus y) = \neg x \odot \neg y$. Ax. 6'. $\neg(x \odot y) = \neg x \oplus \neg y$.
Ax. 7. $x = \neg\neg x$. Ax. 8. $\neg 0 = 1$.

In order to write the remaining axioms the following definition is given: $x \vee y = (x \odot \neg y) \oplus y$, $x \wedge y = (x \oplus \neg y) \odot y$.

Ax. 9. $x \vee y = y \vee x$. Ax. 9'. $x \wedge y = y \wedge x$.
Ax. 10. $x \vee (y \vee z) = (x \vee y) \vee z$. Ax. 10'. $x \wedge (y \wedge z) = (x \wedge y) \wedge z$.
Ax. 11. $x \oplus (y \wedge z) = (x \oplus y) \wedge (x \oplus z)$. Ax. 11'. $x \odot (y \wedge z) = (x \odot y) \wedge (x \odot z)$.

This definition is equivalent to the definition presented in Basic Notions on Universal algebras subsection.

With respect to the operations $\oplus$, $\odot$, and $\neg$ the distinguishing feature between an MV-algebra $(A, \oplus, \odot, \neg, 0, 1)$ and a Boolean algebra is the lack of the idempotent law $x \oplus x = x$, whereas with respect to the operations $\vee$, $\wedge$, and $\neg$ the difference between the system $(A, \vee, \wedge, \neg, 0, 1)$ and a Boolean algebra is the lack of the law of the excluded middle $x \vee \neg x = 1$.

A lattice-ordered abelian group (ℓ-group) is an algebra $(G, +, -, 0, \vee, \wedge)$ such that $(G, +, -, 0)$ is an abelian group, $(G, \vee, \wedge)$ is a lattice, and $+$ distributes over $\vee$

A. Di Nola et al., *Fuzzy Logic of Quasi-Truth: An Algebraic Treatment*,
Studies in Fuzziness and Soft Computing 338, DOI 10.1007/978-3-319-30406-9_4

and $\wedge$. A totally-ordered abelian group (o-group) is an ℓ-group in which the order is total. A *strong unit* of the ℓ-group G is an element $u > 0$ of G such that, for every $a \in G$, there exists a natural number m with $a \leq mu$.

Let (G, u) be an ℓ-group equipped with a fixed strong unit u. $\Gamma(G, u)$ is the structure $\Gamma(G, u) = ([0, u], \oplus, \neg, 0)$ defined as follows:

$[0, u] = \{a \in G : 0 \leq a \leq u\}$

$a \oplus b = (a + b) \wedge u$

$\neg a = u - a$

0 = the additive identity 0 of G.

$\Gamma(G, u)$ is an MV-algebra. The construction of $\Gamma(G, u)$ from (G, u) is due to Chang [2] for the totally-ordered case, and to Mundici [3] for the general case. We have the following

Proposition 4.1 [3] *(i) the lattice-order induced by the MV-algebra operations in $\Gamma(G, u)$ coincides with the order inherited from G;*

(ii) if $h : (G_1, u_1) \to (G_2, u_2)$ is an ℓ-group homomorphism mapping u_1 to u_2, then the restriction Γh of h to $[0, u_1]$ is an MV-algebra homomorphism $\Gamma h : \Gamma(G_1, u_1) \to (G_2, u_2)$;

(iii) Γ is a full, faithful, and dense functor (i.e., a categorical equivalence) between the category of ℓ-groups with strong unit and the category of MV-algebras. In particular, for every MV-algebra A, there exists a unique ℓ-group with strong unit (G, u) such that A is isomorphic to $\Gamma(G, u)$. If A is countable, then A is countable;

(vi) the ideals (i.e., kernels of homomorphisms) of (G, u) correspond bijectively to the ideals of $\Gamma(G, u)$ via the inclusion-preserving application $\mathcal{I} \mapsto \mathcal{I} \cap [0, u]$, whose inverse is $I \mapsto$ (ideal generated by I in G). If $I = \mathcal{I} \cap [0, u]$, then $\Gamma(G, u)/I$ and $\Gamma(G/\mathcal{I}, u/\mathcal{I})$ are isomorphic via $a/I \mapsto a/\mathcal{I}$.

Let A be an MV-algebra. For any $x, y \in A$ we write $x \leq y$ iff $\neg x \oplus y = 1$. Then, as proved by Chang [1], $\leq$ induces a partial order relation. Specifically, the order endows A with a bounded distributive lattice structure, where the join $x \vee y$ and the meet $x \wedge y$ are given by $x \vee y = \neg(\neg x \oplus y) \oplus y$ and $x \wedge y = \neg(\neg x \vee \neg y)$.

4.2 Examples of MV-Algebras

The first and most important example of an MV-algebra is the Lindenbaum algebra $(\mathbf{Form}(\text{Ł})/\equiv, \to, \neg)$ obtained from Łukasiewicz propositional calculus Ł where $\alpha \equiv \beta$ iff $\vdash_{\text{Ł}} \alpha \leftrightarrow \beta$ for any $\alpha, \beta \in \mathbf{Form}(\text{Ł})$ [1, 3].

The unit interval of real numbers $[0, 1]$ endowed with the following operations: $x \oplus y = \min(1, x + y)$, $x \odot y = \max(0, x + y - 1)$, $\neg x = 1 - x$, becomes an MV-algebra. It is well known that the MV-algebra $S = ([0, 1], \oplus, \odot, \neg, 0, 1)$ generate the variety $\mathbf{MV}$ of all MV-algebras, i.e. $\mathcal{V}(S) = \mathbf{MV}$.

Following [4], the MV-algebras S_m and S_m^ω , for $m \geq 1$, are defined as follows: $S_m = \Gamma(Z, m)$ $S_m^\omega = \Gamma(Z \times_{lex} Z, (m, 0))$, where $Z \times_{lex} Z$ is the lexicographic product of two copies of the o-group Z of the integers.

Chang's MV-algebra C [1], which is our main interest, is defined on the set

$$C = \{0, c, ..., nc, ..., 1 - nc, ..., 1 - c, 1\}$$

by the following operations (consider $0 = 0c$): $x \oplus y =$

- $(m + n))c$ if $x = nc$ and $y = mc$
- $1 - (m - n)c$ if $x = 1 - nc$ and $y = mc$ and $0 < n < m$
- $1 - (n - m)c$ if $x = nc$ and $y = 1 - mc$ and $0 < m < n$
- 1 otherwise;

$\neg x = 1 - nc$ if $x = nc$, $\neg x = nc$ if $x = 1 - nc$.

The MV-algebra C is isomorphic to the algebra S_1^{ω}. Last but not least, we construct Chang's MV-algebra in a way which reflects the logical structure related to a Pavelka style fuzzy logic, to be studied in Chap. 11; this construction is also easily visualized. Recall [5] a *Product algebra* P is a BL-algebra which satisfies additional conditions

$$x^{**} \leq (y \odot x \to z \odot x) \to (y \odot z),$$
$$x \wedge x^* = \mathbf{0}$$

for all $x, y, z \in P$, where x^* stands for $\neg x$ (another commonly used notation for complement). A simple example is based on the product t-norm $\odot$ on the real unit interval $[0, 1]$; $x \odot y = xy$.

Fix an element $t \in P, 0 < t < 1$. Then the set $T = \{t^n \mid n \geq 0\}$ is an infinite decreasing chain

$$\cdots < t^n < \cdots < t^3 < t^2 < t < t^0 = \mathbf{1}.$$

In fact T is a cancellative lattice-ordered monoid. Now reverse the order and rename the elements t^n by f^n as follows

$$\mathbf{0} = f^0 < f < f^2 < f^3 < \cdots < f^n < \cdots$$

Then the set $F = \{f^n \mid n \geq 0\}$ is an infinite increasing chain. Assuming $f^n < t^n$ for any natural $n \geq 0$, we construct the set $F \cup T$

$$\mathbf{0} < f < f^2 < f^3 < \cdots < f^n < \cdots \quad \cdots < t^n < \cdots < t^3 < t^2 < t < \mathbf{1}.$$

(Here the superscripts of t, f only index these elements, they do not mean any type of power, repeated multiplication or $\odot$-operation). Notice that $F \cap T = \emptyset$ and $F \cup T$ is a lattice that is not complete as $\bigvee F$ and $\bigwedge T$ do not exist in $F \cup T$; however, if a supremum of a subset of the set $F \cup T$ exists, then it is the greatest element of this subset (and conversely). Similarly, if an infimum of a subset of the set $F \cup T$ exists, then it is the smallest element of this subset (and conversely). We now define the operations $\oplus$ and * on $F \cup T$ as follows: for any $m, n \geq 0$, $(f^n)^* = t^n$, $(t^n)^* = f^n$.

Moreover,

$$f^m \oplus f^n = f^{m+n},$$
$$t^m \oplus t^n = \mathbf{1},$$
$$f^m \oplus t^n = \begin{cases} t^{n-m} & \text{if } n > m, \\ \mathbf{1} & \text{otherwise.} \end{cases}$$

The product operation $\odot$ obeys dual equations

$$t^m \odot t^n = t^{m+n},$$
$$f^m \odot f^n = \mathbf{0},$$
$$t^m \odot f^n = \begin{cases} f^{n-m} & \text{if } n > m, \\ \mathbf{0} & \text{otherwise.} \end{cases}$$

It is a routine task to show that by setting $C = F \cup T$ we obtain an MV-algebra that is isomorphic to Chang's MV-algebra. The MV-algebra C is a prototypical example of a *perfect* MV-algebra; any element $c \in C$ satisfies the equation

$$(c \oplus c) \odot (c \oplus c) = (c \odot c) \oplus (c \odot c). \tag{4.1}$$

4.3 Properties of MV-Algebras

In this subsection we give some identities which are consequence of MV-algebra axioms.

Proposition 4.2 [1] *(i)* $x \vee 0 = x = x \wedge 1,\ x \wedge 0 = 0, x \vee 1 = 1.$
(ii) $x \vee x = x = x \wedge x.$
(iii) $\neg(x \vee y) = \neg x \wedge \neg y,\ \neg(x \wedge y) = \neg x \vee \neg y.$
(iv) $x \wedge (x \vee y) = x = x \vee (x \wedge y).$
(v) If $x \oplus y = 0$, *then* $x = y = 0.$
(vi) If $x \odot y = l$, *then* $x = y = l.$
(vii) If $x \vee y = 0$, *then* $x = y = 0.$
(viii) If $x \wedge y = 1$. *then* $x = y = 1.$

Proposition 4.3 [1] *Let B be the set of elements $x \in A$ such that $x \oplus x = x$. Then B is closed under the operations $\oplus$, $\odot$, and $\neg$ where $x \oplus y = x \vee y$ and $x \odot y = x \wedge y$ for $x, y \in B$. Furthermore, the system $(B, \oplus, \odot, \neg, 0, 1)$ is not only a subalgebra of A but is also the largest subalgebra of A which is at the same time a Boolean algebra with respect to the same operations $\oplus$, $\odot$, and $\neg$.*

By definition (i) $0x = 0$ and $(n+1)x = nx \oplus x$. (ii) $x^0 = 1$ and $x^{n+1} = (x^n) \odot x$. The *order* of an element x, in symbols $ord(x)$, is the least integer m such that $mx = 1$. If no such integer m exists then $ord(x) = \infty$.

Proposition 4.4 [1] *(i) If $x \vee y = 1$, then $x^n \vee y^n = 1$ for each n.*
(ii) If $ord(x \odot y) < \infty$, then $x \oplus y = 1$.
(iii) If $ord(x) > 2$, then $ord(x \odot x) = \infty$.

An MV-algebra A is *simple* if, and only if, every element of A different from 0 has a finite order. An MV-algebra A is linearly ordered if, and only if, for every $x, y \in A$, either $x \leq y$ or $y \leq x$.

Proposition 4.5 [1] *(i) Every simple MV-algebra is linearly ordered.*
(ii) If A is linearly ordered, then $x \oplus z = y \oplus z$ and $x \oplus z \neq 1$ implies $x = y$.

Remark 4.6 If $a \neq \mathbf{1}, b \neq \mathbf{0}$ are elements of a linearly ordered MV-algebra (in particular, Chang's MV-algebra), then $a \oplus b > a$.

4.4 Ideals, Filters, Congruence Relations

A subset I of an MV-algebra A is an *ideal* of A if, and only if, (i) $0 \in I$, (ii) if $x, y \in I$, then $x \oplus y \in I$, and (iii) if $x \in I$ and $y \leq x$, then $y \in I$. An ideal I is said to be *proper* if $I \neq A$. Clearly an ideal I is proper if, and only if, $1 \notin I$.

Dually, a subset F of an MV-algebra A is a *filter* of A if, and only if, (i) $1 \in F$, (ii) if $x, y \in F$, then $x \odot y \in F$, and (iii) if $x \in F$ and $x \leq y$, then $y \in F$. A filter F is said to be *proper* if $F \neq A$. Clearly a filter F is proper if, and only if, $0 \notin F$.

Let us denote by $Spec A$ the set of all prime ideals of A. As it is well known, $Spec A$ equipped with set-theoretical inclusion is a root system.

Proposition 4.7 [1] *(i) If f is a homomorphism of an MV-algebra A onto another MV-algebra, then the set of elements $x \in A$ such that $f(x) = 0$ ($f(x) = 1$) is an ideal (a filter) and the relation E defined by xEy if and only if $f(x) = f(y)$ is a congruence relation.*

(ii) If E is a congruence relation, then the set of elements of $0/E$ $(1/E)$ is an ideal (a filter).

(iii) If E is a congruence relation, then xEy if and only if $(\neg x \odot y) \oplus (\neg y \odot x) E 0$ $((\neg x \oplus y) \odot (\neg y \oplus x) E 0)$.

(iv) If E_1 and E_2 are congruence relations, then $E_1 = E_2$ if and only if $0/E_1 = 0/E_2$ $(1/E_1 = 1/E_2)$.

(v) If I (F) is an ideal (a filter), then the relation E defined by xEy if and only if $(\neg x \odot y) \oplus (\neg y \odot x) \in I$ $((\neg x \oplus y) \odot (\neg y \oplus x) \in F)$ is a congruence relation. So, there exists one-to-one correspondence between the set of ideals (filters) and the set of congruences: if E is a congruence of the MV-algebra A, then $E \mapsto \{x \in A : xE0\}$ $(E \mapsto \{x \in A : xE1\})$; is I (F) is an ideal (filter) of A, then $I \mapsto \{(x, y) \in A^2 : (\neg x \odot y) \oplus (\neg y \odot x) \in I\}$ $(F \mapsto \{(x, y) \in A^2 : (\neg x \oplus y) \odot (\neg y \oplus x) \in F\})$. The equivalence class a/E we will denote as a/I (a/F) or $\frac{a}{I}$ (or $\frac{a}{F}$, where I (F) is the ideal (filter) corresponding to the congruence relation E.

M is a maximal ideal (filter) of A if, and only if, M is a proper ideal (filter) and whenever I (F) is an ideal (a filter) such that $M \subseteq I \subseteq A$ $(M \subseteq F \subseteq A)$, then either $M = I$ or $I = A$ $(M = F$ or $F = A)$.

We say that P is a *prime* ideal (filter) of an MV-algebra A if, and only if, (i) P is an ideal (a filter) of A, and (ii) for each $x, y \in A$, either $\neg x \odot y \in P$ $(\neg x \oplus y \in P)$ or $x \odot \neg y \in P$ $(x \oplus \neg y \in P)$. An ideal H of an MV-algebra A is called *primary* iff $a \odot b \in H$ implies $a^n \in H$ or $b^n \in H$ for some integer n.

Proposition 4.8 [2]

If P is a prime ideal (filter) of A, then A/P is a linearly ordered MV-algebra. If M is a maximal ideal (filter) of A, then A/M is a simple MV-algebra.

Proposition 4.9 [2] *Every MV-algebra is a subdirect product of linearly ordered MV-algebras.*

For any MV-algebra A, the *radical* of A, denoted by $Rad(A)$, is the intersection of all maximal ideals of A.

Non zero elements of $Rad(A)$ are called *infinitesimal*, indeed, $x \in Rad(A)$ if and only if for every $n \in \mathbb{N}$, $nx < \neg x$. If $x \in Rad(A)$ then $x \odot x = 0$ [6] and $ord(\neg x) = 2$.

Theorem 4.10 [7] *Up to isomorphism, every MV-algebra A is an algebra of* $[0, 1]^*$-*valued functions over $Spec(A)$, where* $[0, 1]^*$ *is a ultrapower on the MV-algebra* $[0, 1]$, *depending only on the cardinality of A.*

References

1. Chang, C.C.: Algebraic analysis of many-valued logics. Trans. Am. Math. Soc. **88**, 476–490 (1958)
2. Chang, C.C.: A new proof of the completeness of the Lukasiewicz axioms. Trans. Am. Math. Soc. **93**, 74–80 (1959)
3. Mundici, D.: Interpretation of AF C^*-algebras in Łukasiewicz sentential calculus. J. Funct. Anal. **65**, 15–63 (1986)
4. Komori, Y.: Super-Łukasiewicz propositional logic. Nagoya Math. J. **84**, 119–133 (1981)
5. Hájek, P.: Metamathematics of Fuzzy Logic. Kluwer, Dordrecht (1998)
6. Belluce, L.P.: Semisimple algebras of in infinite-valued logic and bold fuzzy set theory. Canad. J. Math. **38**, 1356–1379 (1986)
7. Di Nola, A.: Representation and reticulation by quotients of MV-algebras. Ricerche di Matematica **40**, 291–297 (1991)

Chapter 5
Local MV-Algebras

Local MV-algebras are MV-algebras with only one maximal ideal that, hence, contains all infinitesimal elements. This class of algebras contains MV-chains and perfect MV-algebras.

An MV-algebra A is called *local* if it has only one maximal ideal, coinciding with $Rad(A)$. If A is a local MV-algebra then $A/Rad(A)$ is a simple MV-algebra, since it does not have non-trivial ideals. We denote simply by $\equiv$ the equivalence $\equiv_{Rad(A)}$.

Proposition 5.1 *An MV-algebra A is local if and only if for every $x \in A$, either $ord(x) < \infty$ or $ord(\neg x) < \infty$.*

Proof Let A be a local MV-algebra and M the maximal ideal of A. Then for every $x \in A$, $ord(x) = \infty$ implies that $x \in M$. Hence, if we assume that $ord(a) = ord(\neg a) = \infty$ for every $a \in A$, then $a, \neg a \in M$, which is impossible. In consequence, $ord(x) < \infty$ or $ord(\neg x) < \infty$, for every $x \in A$. Viceversa, assume that for every $x \in A$, either $ord(x) < \infty$ or $ord(\neg x) < \infty$. Let M be a maximal ideal of A. Suppose $a \notin M$ for some a with $ord(a) = \infty$. Then for some n $(\neg a)^n \in M$. Thus $ord((\neg a)^n) = \infty$ and $ord(na) < \infty$. So $ord(a) < \infty$, which is impossible. Hence every element of infinite order belongs to M, so M is a unique maximal ideal and A is local. □

Proposition 5.2 *Let A be an MV-algebra and H an ideal of A. Then $\frac{A}{H}$ is local if and only if H is primary.*

Proof Suppose that $\frac{A}{H}$ is local and that $a \odot b \in H$. Assume for all n, $a^n \notin H$. Now $\frac{a}{H} \odot \frac{b}{H} = \frac{a \odot b}{H} = 0$, thus $\frac{a}{H} \leq \frac{\neg b}{H}$. For all n, $(\frac{a}{H})^n \neq 0$, thus $n(\frac{\neg a}{H}) \neq 1$. Since $\frac{A}{H}$ is local, it follows that, for some m, $m(\frac{a}{H}) = 1$. Hence $m(\frac{\neg b}{H}) = 1$ and so $\frac{b^m}{H} = 0$, i.e., $b^m \in H$. Thus H is primary. Conversely, suppose H is primary. Let $\frac{a}{H} \in \frac{A}{H}$. Since $a \odot \neg a \in H$, we know $a^n \in H$ or $(\neg a)^n \in H$ for some n. Thus $(\frac{a}{H})^n = 0$ or $(\frac{\neg a}{H})^n = 0$, which implies $n(\frac{\neg a}{H}) = 1$ or $n(\frac{a}{H}) = 1$. Hence $\frac{A}{H}$ is local. □

A. Di Nola et al., *Fuzzy Logic of Quasi-Truth: An Algebraic Treatment*,
Studies in Fuzziness and Soft Computing 338, DOI 10.1007/978-3-319-30406-9_5

Proposition 5.3 *Let A be a local MV-algebra. Then for every $a \in A$ and for every $P, Q \in Spec(A)$ (= the set of all prime ideals of A), we have*

$$\frac{\frac{a}{P}}{Rad(\frac{A}{P})} = \frac{\frac{a}{Q}}{Rad(\frac{A}{Q})}$$

Proof For every $P \in Spec(A)$, $\frac{\frac{A}{P}}{Rad(\frac{A}{P})}$ is simple, and then, up to isomorphism, is a subalgebra of $[0, 1]$. Now, by contradiction, assume that there are $r, s \in [0, 1]$, with $r < s$, such that $r = \frac{\frac{a}{P}}{Rad(\frac{A}{P})}$ and $r = \frac{\frac{a}{P}}{Rad(\frac{A}{P})}$. Then there is a simple term ϕ such that $\phi(r) = 0$ and $\phi(s) = 1$. Hence

$$\phi(\frac{\frac{a}{P}}{Rad(\frac{A}{P})}) = \frac{\frac{\phi(a)}{P}}{Rad(\frac{A}{P})} = 0,$$

thus, $\frac{\phi(a)}{P} \in Rad(\frac{A}{P})$ while

$$\phi(\frac{\frac{a}{Q}}{Rad(\frac{A}{Q})}) = \frac{\frac{\phi(a)}{Q}}{Rad(\frac{A}{Q})} = 1$$

hence, $\frac{\neg\phi(a)}{Q} \in Rad(\frac{A}{Q})$. Thus, $ord(\phi(a)) = \infty$ and $ord(\neg\phi(a)) = \infty$, in contrast with the assumption that A is local. □

Theorem 5.4 *The class of all local MV-algebras is a universal class.*

Proof Let A be a local MV-algebra. We claim that the following statement holds:

$$For\ \ every\ \ x \in A, x \leq \neg x,\ \ or\ \ \neg x \leq x\ \ or\ \ (d(x, \neg x))^2 = 0, \qquad (\dagger)$$

where $d(x, y) = (\neg x \odot y) \oplus (x \odot \neg y)$.

Indeed for every $x \in A$, if $x \equiv \neg x$ does not hold, we have either $x < \neg x$ or $\neg x < x$. In the case that $x \equiv \neg x$ we have $d(x, \neg x) \in Rad(A)$, then $(d(x, \neg x))^2 = 0$. Hence ($\dagger$) holds. Assume now that ($\dagger$) holds. If $x \leq \neg x$ then $ord(\neg x) < \infty$. Analogously for $\neg x < x$, then $ord(x) < \infty$. If $(d(x, \neg x))^2 = 0$, i.e., $x^2 \oplus (\neg x)^2)^2 = 0$, then for every prime ideal P of A we have the following cases:

(i) $\frac{x}{P} \leq \frac{\neg x}{P}$;
(ii) $\frac{\neg x}{P} \leq \frac{x}{P}$.

Assuming (i), we get $\frac{x}{P} \odot \frac{x}{P} = 0$, and then $(\frac{\neg x}{P})^4 = 0$. Hence $ord(\frac{x}{P}) \leq 4$. While, assuming (ii), we get $ord(x) \leq 2$. Hence, for every prime ideal P of A, $ord(\frac{x}{P}) \leq 4$. This implies that $ord(x) < \infty$. Hence A is local. □

Theorem 5.5 *Every MV-algebra has a greatest local subalgebra.*

Proof Let A be an MV-algebra. Then for every prime ideal P of A, the algebra $\frac{\frac{x}{P}}{Rad(\frac{A}{P})}$ is simple, hence it is isomorphic to a subalgebra of [0, 1]. Let

$$\mathcal{L}(A) = \{x \in A \mid for \quad every \quad P \in Spec(A), \quad \frac{\frac{x}{P}}{Rad(\frac{A}{P})} = r_x \in [0, 1]\}.$$

We can easily check that $\mathcal{L}(A)$ is a subalgebra of A. Let us prove that $\mathcal{L}(A)$ is local. Let $y \in \mathcal{L}(A)$ such that $\frac{\frac{y}{P}}{Rad(\frac{A}{P})} = 0$ for all $P \in Spec(A)$. This is equivalent to $\frac{y}{P} \in Rad(\frac{A}{P})$, and $\frac{\neg y}{P} \in \neg Rad(\frac{A}{P})$. Hence $ord(\neg y) < \infty$. Assume now that $\frac{\frac{y}{P}}{Rad(\frac{A}{P})} = r \in (0, 1]$, for all $P \in Spec(A)$. Then there exists n such that $nr = 1$, so $\frac{\frac{ny}{P}}{Rad(\frac{A}{P})} = 1$ for all $P \in Spec(A)$. This is equivalent to say that for every prime ideal P, $\frac{\neg y}{P} \in Rad(\frac{A}{P})$. Hence $ord(ny) < \infty$ and so $ord(y) < \infty$. Hence $\mathcal{L}(A)$ is local. To show that $\mathcal{L}(A)$ is the greatest local subalgebra of A, let B be a local subalgebra of A and suppose that there is an element $b \in B \backslash \mathcal{L}(A)$, i.e.,

$$r_b^P = (\frac{\frac{b}{P}}{Rad(\frac{A}{P})}) \neq (\frac{\frac{b}{Q}}{Rad(\frac{A}{Q})}) = r_b^Q$$

for $P, Q \in Spec(A)$, where $r_b^P, r_b^Q \in [0, 1]$. Then there is an MV-term f such that $f(r_b^P) = 0$ and $f(r_b^Q) = 1$. It is easy to check that $ord(f(r_b^P)) = \infty$ and $ord(\neg f(r_b^Q)) = \infty$. But this implies that $ord(f(b)) = \infty$ and $ord(\neg f(b)) = \infty$, which is equivalent to $ord(b) = \infty$ and $ord(ny) = \infty$. This is in contradiction with the assumption of B being local. □

Let us give a class of examples of local MV-algebra. Let X be an arbitrary non empty set, A an MV-algebra, and $K(A^X)$ the subset of the MV-algebra A^X defined as follows:

$$K(A^X) = \{f \in A^X : f(X) \subseteq a/Rad(A) \text{ for some } a \in A\}.$$

Proposition 5.6 [1] *$K(A^X)$ is a local MV-algebra.*

Proof The zero constant function f_0 belongs to $K(A^X)$, in fact $f_0(X) = \{0\} \subseteq Rad(A) = \frac{0}{Rad(A)}$. Assume f satisfies $f(X) \subseteq \frac{a}{Rad(A)}$ for some $a \in A$. Then, $\neg f(X) \subseteq \frac{a}{Rad(A)}$. Finally, let $f, g \in K(A^X)$ be such that $f(X) \subseteq \frac{a}{Rad(A)}$ for some $a \in A$ and $g(X) \subseteq \frac{b}{Rad(A)}$ for some $b \in A$. Then $(f \oplus g)(X) \subseteq \frac{(a \oplus b)}{Rad(A)}$. Hence $K(A^X)$ is a subalgebra of A. Let us show that $K(A^X)$ is local. Take $f \in K(A^X)$. If $f(X) \subseteq \frac{0}{Rad(A)}$ then $\neg f(X) \subseteq \frac{1}{Rad(A)}$ and $ord(\neg f) < \infty$. If $f(X) \subseteq \frac{1}{Rad(A)}$ then $ord(f) < \infty$. Now, assume that $\neg f(X) \subseteq \frac{a}{Rad(A)} \neq \frac{0}{Rad(A)} \neq \frac{1}{Rad(A)}$. Then for every $x \in X$, $f(x) \equiv_{Rad(A)} a$ and $a \notin Rad(A)$. Since $\frac{A}{Rad(A)}$ is an MV-chain, we have $ord(f) < \infty$ and $ord(\neg f) < \infty$. □

Any element f of an algebra $K(A^X)$ from the above class will be called a *quasi constant* function. The algebra $K(A^X)$ will be called *the full MV-algebra of quasi constant functions* from X to A. Using full MV-algebras of quasi constant functions a representation theorem for all local MV-algebras can be obtained, as we show below.

Theorem 5.7 *Every local MV-algebra can be embedded into a full MV-algebra of quasi constant functions.*

Proof Let A be a local MV-algebra. Any element $x \in A$ is a function from $Spec(A)$ into $[0, 1]^*$, hence for every $P \in Spec(A)$, $\frac{x}{P} \in [0, 1]^*$. For any $x \in A$ there exists $r_x \in [0, 1]$ such that $\frac{\frac{x}{P}}{Rad(\frac{x}{P})} = r_x$. Since every $\frac{A}{P}$ is embeddable in $[0, 1]^*$, $Rad(\frac{A}{P})$ is embeddable in $Rad([0, 1]^*)$. Hence, for every $P \in Spec(A)$, we have $\frac{x}{P} \subseteq \frac{r_x}{Rad([0,1]^*)}$, and so A is an algebra of quasi constant functions. □

Reference

1. Di Nola, A., Esposito, I., Gerla, B.: Local algebras in the representation of MV-algebras. Algebra Universalis **56**, 133–164 (2007)

Chapter 6
Perfect MV-Algebras

The aim of this section is to give an account of the class of Perfect MV-algebras. Such a class is a full subcategory of the category of MV-algebras. In general, there are MV-algebras which are not semisimple. Roughly speaking we can say that a non semisimple MV-algebra A has non-zero radical. We call a non-zero element from the radical of A an infinitesimal. A first example of non simple MV-chain was given by Chang in [1], where the MV-algebra C is described. The algebra C has remarkable properties that we will try to display through the following chapters. Indeed it is easy to check that:

(1) C is generated by its radical
(2) $C = Rad(C) \cup \neg Rad(C)$
(3) $C/Rad(C) \cong \{0, 1\}$.

Hence C is just made by infinitesimal elements and co-infinitesimal elements. We then would like to describe a class of MV-algebras containing C and whose elements share the above properties. Then we can think of such a class as the one made by MV-algebras which are, up to infinitesimal elements, like the 2-elements Boolean algebra $\{0, 1\}$. We say that an MV-algebra A is *perfect* if for each element $x \in A$, $ord(x) < \infty$ iff $ord(\neg x) = \infty$.

Proposition 6.1 *Let A be a perfect MV-algebra. Then $Rad(A)$ is the unique maximal ideal of A.*

Proof It is clear that $Rad(A)$ is an ideal. Let $x, y \in A$ such that $x \wedge y \in Rad(A)$. Assume that $x, y \notin Rad(A)$, then $ord(x) < \infty$ and $ord(y) < \infty$. Hence $2x = 1$, $2y = 1$ and $2(x \wedge y) = 1$. That is $ord(x \wedge y) = 2$, in contradiction with $x \wedge y \in Rad(A)$. So either $x \in Rad(A)$ or $y \in Rad(A)$. Since any prime ideal of A cannot contain any element of finite order, then we get that $Rd(A)$ is the unique maximal ideal of A. □

We say that an ideal J of an MV-algebra A is *perfect* if for every $x \in A$, there is an $n \in \mathbb{N}$ such that $x^n \in J$ iff $(\neg x)^m \notin J$ for all $m \in \mathbb{N}$. Then we have

A. Di Nola et al., *Fuzzy Logic of Quasi-Truth: An Algebraic Treatment*,
Studies in Fuzziness and Soft Computing 338, DOI 10.1007/978-3-319-30406-9_6

Proposition 6.2 [2] *An ideal $J \subseteq A$ is perfect iff A/J is perfect.*

Proposition 6.3 *Let A be an MV-algebra, the following statements are equivalent:*

(1) A is perfect,
(2) every ideal $J \subseteq A$ is perfect.

Proof If all ideals of A are perfect, then $\{0\}$ ideal is perfect, hence $A \cong A/\{0\}$ is perfect. Conversely, let A be perfect and J an ideal of A. Assume $x \in A$ and $x^n \in J$. Then $ord(x^n) = \infty$ so $ord(n(\neg x)) < \infty$ and then $ord(\neg x) < \infty$. If for some m, $(\neg x)^m \in J$, we similarly have $ord(x) < \infty$, which is impossible. So for no m do we have $(\neg x)^m \in J$. Assume now that $(\neg x)^m \notin J$ for any m. Then $(\neg x)^m \neq 0$ for any m, so $mx \neq 1$, for any m. Thus $ord(x) = \infty$ and $ord(\neg x) < \infty$. Hence $n(\neg x) = 1$ for some n, so $x^n = 0 \in J$. So J is perfect. □

For an MV-algebra A and an ideal I in A, let $\langle I \rangle$ denote the subalgebra generated by I. Then $\langle I \rangle = I \cup \neg(I)$ where $\neg(I) = \{x \mid \neg x \in I\}$.

Proposition 6.4 *In an MV-algebra A, $\langle Rad(A) \rangle$ is a perfect subalgebra of A.*

Proof Let $x \in \langle Rad(A) \rangle$. If $x \in Rad(A)$, then $x^2 = 0$ and $ord(x) = \infty$. So $2(\neg x) = 1$ and $ord(\neg x) < \infty$. If $x \in \neg(Rad(A))$ then $\neg x \in Rad(A)$. So $ord(\neg x) = \infty$ and $ord(x) < \infty$. □

Proposition 6.5 *Let A be a perfect MV-algebra. Then $A = \langle Rad(A) \rangle$.*

Proof Clearly $\langle Rad(A) \rangle$ is a subalgebra of A. Since A is perfect, hence $Rad(A)$ is the unique maximal ideal of A and consists of all elements of infinite order. Thus if $x \in A$ and $ord(x) = \infty$, then $x \in Rad(A)$. If $x \in A$ and $ord(x) < \infty$ then $ord(\neg x) = \infty$ so $\neg x \in \neg(Rad(A))$. Thus $A \subseteq \langle Rad(A) \rangle$. □

Proposition 6.6 *Let A be an MV-algebra. Then the following are equivalent:*

(1) A is perfect,
(2) $A/Rad(A) = \{0, 1\}$.

Proof Let A be perfect. Then the ideal $Rad(A)$ is perfect and maximal. Thus $A/Rad(A)$ is perfect and simple. Hence $A/Rad(A) = \{0, 1\}$. On the other hand assume $A/Rad(A) = \{0, 1\}$. Let $x \in A$, then $x/Rad(A) = 0$ or $x/Rad(A) = 1$. That is $x \in Rad(A)$ or $x \in \neg(Rad(A))$. So $A = \langle Rad(A) \rangle$, hence A is perfect. □

Proposition 6.7 *Let A be a perfect MV-algebra and f a homomorphism to an MV-algebra. Then $f(A)$ is a perfect MV-algebra.*

Proof Let A be perfect. Then $A = Rad(A) \cup \neg(Rad(A))$. Let $x \in Rad(A)$. Then for every integer $n \geq 0$ we have $nx \leq \neg x$, which implies $nf(x) \leq \neg f(x)$. Hence, $f(x) \in Rad(f(A))$ and $f(Rad(A)) \subseteq Rad(f(A))$. If $x \in \neg Rad(A)$, then, for every integer $n \geq 0$, $n(\neg x \leq x$, and $n(\neg f(x)) \leq f(x)$. So we get $f(x) \in \neg Rad(f(A))$ and $f(\neg Rad(A)) \subseteq \neg Rad(f(A))$. Then we have that $f(A) = f(Rad(A)) \cup f(\neg Rad(A)) \subseteq Rad(f(A)) \cup \neg Rad(f(A)) \subseteq f(A)$. Thus $f(A) = Rad(f(A)) \cup \neg Rad(f(A))$, that is $f(A)$ is perfect. □

Let A be an MV-algebra and P a perfect subalgebra of A. Then $P = \langle Rad(P) \rangle$. Now $Rad(P) = P \cap Rad(A)$ so we see that $\langle Rad(P) \rangle \subseteq \langle Rad(A) \rangle$. Hence $\langle Rad(A) \rangle$ is a perfect subalgebra of A that contains all perfect subalgebras of A. Call such perfect subalgebra of A the *perfect skeleton* of A, denoted by Perf(A).

Proposition 6.8 *Let A be a non semisimple MV-algebra. Then A contains as a copy of C as subalgebra, actually as a subalgebra of Perf(A).*

Proof Since A is non semisimple, then $Rad(A) \neq \{0\}$. Let z be a non zero element of $Rad(A)$ and $id(z)$ denote the ideal of A generated by z. Let ϕ be a map from C to $\langle id(z) \rangle$ defined as follows: for every $n \in \mathbb{N}$, $\phi(nc) = nz$ and $\phi(\neg nc) = \neg nz$. It is easy to check that ϕ is an isomorphism between C and $\langle id(z) \rangle$. □

For every MV-algebra A the *perfect radical ideal* of A is the ideal

$$\sqrt{}_p(A) = \bigcap \{J \mid J \text{ is a perfect ideal of } A\}.$$

Proposition 6.9 *Let A be an MV-algebra. Then*

$$\sqrt{}_p\left(\frac{A}{\sqrt{}_p(A)}\right) = 0.$$

Proof If A has no perfect ideals then $\sqrt{}_p(A) = A$. Let $I/\sqrt{}_p(A)$ be a perfect ideal in $\frac{A}{\sqrt{}_p(A)}$. Consider the map

$$\frac{\left(\frac{A}{\sqrt{}_p(A)}\right)}{\frac{I}{\sqrt{}_p(A)}} \to \frac{A}{I},$$

with $\sqrt{}_p(A) \subseteq I$, given by

$$\frac{\frac{x}{\sqrt{}_p(A)}}{\frac{I}{\sqrt{}_p(A)}} \to \frac{x}{I}.$$

In order to prove that the above map is well-defined suppose that

$$\frac{\frac{x}{\sqrt{}_p(A)}}{\frac{I}{\sqrt{}_p(A)}} = \frac{\frac{y}{\sqrt{}_p(A)}}{\frac{I}{\sqrt{}_p(A)}}$$

so that $d(\frac{x}{\sqrt{}_p(A)}, (\frac{y}{\sqrt{}_p(A)}) \in \frac{I}{\sqrt{}_p(A)}$. Then $\frac{d(x,y)}{\sqrt{}_p(A)} \in \frac{I}{\sqrt{}_p(A)}$ and since $\sqrt{}_p(A) \subseteq I$ we have $d(x, y) \in I$ so the map is well defined. It is easy to check that the map is an epimorphism. As epimorphic image of perfect MV-algebras are perfect we see that $\frac{I}{\sqrt{}_p(A)}$ is perfect in $\frac{A}{\sqrt{}_p(A)}$, then I is perfect in A. □

Proposition 6.10 *Let A be an MV-algebra and set*

$$per(A) = \bigcap\{J \mid J \textit{ is a perfect ideal of } A\}.$$

Then $per(\frac{A}{par(A)}) = 0$.

Proof If A has no perfect ideals then $per(A) = A$. Let $\frac{I}{per(A)}$ be a perfect ideal in $\frac{A}{per(A)}$. Consider the map

$$\frac{(\frac{A}{per(A)})}{(\frac{I}{per(A)})} \to \frac{A}{I}, \qquad with \quad per(A) \subseteq I$$

given by $\frac{(\frac{x}{per(A)})}{(\frac{x}{per(A)})} \to \frac{A}{I}$. In order to prove that the above map is well-defined suppose that

$$\frac{(\frac{x}{per(A)})}{(\frac{x}{per(A)})} \to \frac{A}{I} = \frac{(\frac{y}{per(A)})}{(\frac{y}{per(A)})} \to \frac{A}{I}$$

so that $d(\frac{x}{per(A)}, \frac{y}{per(A)}) \in \frac{I}{per(A)}$. Then $\frac{d(x,y)}{per(A)} \in \frac{I}{per(A)}$ and since $per(A) \subseteq I$ we have $d(x, y) \in I$. So the map is well-defined. It is easy to check that the map is an epimorphism. As epimorphic images of perfect MV-algebras are perfect we see that if $\frac{I}{per(A)}$ is perfect in $\frac{A}{per(A)}$, then I is perfect in A. □

Call an MV-algebra A *semi-perfect* if $per(A) = \{0\}$. Thus if $per(A) = \{0\}$, then A is a subdirect product of perfect MV-algebras and then $A \in V(C)$.

6.1 The Category of Perfect MV-Algebras

A relevant fact concerning perfect MV-algebras is that each one of them is generated by its infinitesimals. This turns out to induce a very special structure on the generated algebra. Perfect MV-algebras can be seen as an extreme case of non-archimedean MV-algebras. Thus, the role of perfect MV-algebras is important because it is strictly linked with the role of infinitesimals. An important example of a perfect MV-algebra can be found as a subalgebra S of the Lindenbaum algebra L of First order Łukasiewicz logic. Indeed, the subalgebra S, which is generated by the classes of formulas which are valid but non-provable is a perfect MV-algebra and coincides with Perf(L). Hence perfect MV-algebras are directly connected with a very important phenomenon in Łukasiewicz first order logic, namely, with the incompleteness of such a logic.

Perfect MV-algebras form a full subcategory of the category of all MV-algebras. We denote the category of perfect MV-algebras by **Perfect**.

As it is well known, MV-algebras form a category which is equivalent to the category of abelian lattice ordered groups (l-groups, for short) with strong unit. Let us denote by Γ such an equivalence. This makes the interest in MV-algebras relevant outside the realm of logic. Hence, we know that to each MV-algebra is associated an abelian ℓ-group G with a strong unit, so of course perfect MV-algebras share this property with all MV-algebras. But more, we can functorially map each perfect MV-algebra to an abelian ℓ-group and vice versa, without the help of a strong unit. Let $\mathfrak{A}$ denote the category of abelian ℓ-groups. Let G be an abelian ℓ-group and $G^+ = \{x \in G \mid x > 0\}$ be the positive cone of G. Let $\mathbb{Z} \times_{lex} G$ be the lexicographic product of the additive ℓ-group $\mathbb{Z}$ of integers by G. Give $\mathbb{Z} \times_{lex} G$ the order unit $(1, 0)$; then the MV-algebra $\mathfrak{G}(G) = \Gamma(\mathbb{Z} \times_{lex} G, (1, 0))$ is a perfect MV-algebra. Each element $d \in \mathfrak{G}(G)$ has either the form $d = (0, g)$ for some $g \in G^+ \cup \{0\}$, or $d = (1, g)$ for some $g \in G^- \cup \{0\}$, where $G^- = -G^+$. Thus we got a map $\mathfrak{G}$ from the category of abelian ℓ-groups to the category of perfect MV-algebras, the latter seen as a full subcategory of all MV-algebras. Hence we have the following proposition:

Proposition 6.11 *$\mathfrak{G}$ is a functor from the category $\mathfrak{A}$ to the category* **Perfect**.

Proof Trivial. □

Conversely, now to go back from **Perfect** to $\mathfrak{A}$ let us start with a perfect MV-algebra A. Since $(Rad(A), \oplus, 0)$ is a cancellative monoid, by [3], (Theorem 1, Chapter XIV, Sect. 2), we define the abelian group $\mathcal{D}(A) = ((Rad(A) \times Rad(A)/\nu, \oplus)$, where the binary relation ν is given by $(x, y)\nu(x^{'}, y^{'})$ iff $x \oplus y^{'} = x^{'} \oplus y$ with $x, x^{'}, y, y^{'} \in Rad(A) \times Rad(A)$ and $[x, y] \oplus [x^{'}, y^{'}] = [x \oplus x^{'}, y \oplus y^{'}]$, and $[., .]$ denotes a class of $(Rad(A) \times Rad(A)$ under ν. The neutral element of $\mathcal{D}(A)$ is $[0, 0]$ and the opposite element of $[x, y]$ is $-[x, y] = [y, x]$. The relation $\leq$ define on $\mathcal{D}(A)$ by $[x, y] \leq [x^{'}, y^{'}]$ iff $x \oplus y^{'} \leq x^{'} \oplus y$ turns out to be an order relation.

Proposition 6.12 $(\mathcal{D}(A), \oplus, \leq)$ *is an abelian ℓ-group.*

Proof The proof can be obtained by a direct verification. □

For each MV-homomorphism between perfect MV-algebras $f : A \to A^{'}$ let $\mathcal{D}(f) : \mathcal{D}(A) \to \mathcal{D}(A^{'})$ be defined by $(\mathcal{D}(f))[x, y] = [f(x), f(y)]$.

Theorem 6.13 *$\mathcal{D}$ is a functor from Perfect MV-algebras to the category of abelian l-groups.*

Proof Trivial. □

Theorem 6.14 *The category of Perfect MV-algebras is equivalent to the category of abelian l-groups.*

Proof It can be directly verified that for every $G \in \mathfrak{A}$ and $A \in$ **Perfect** $\mathcal{D}(\mathfrak{G}(G)) \cong G$ and $\mathfrak{G}(\mathcal{D}(A)) \cong A$. Then, in the light of ([4], IV Theorem 1), we get the claimed equivalence. □

Proposition 6.15 *The following statements hold:*

(1) $\{0, 1\}$ *is a terminal and initial object of* **Perfect***;*
(2) **Perfect** *has pull-backs;*
(3) **Perfect** *has arbitrary products;*
(4) **Perfect** *has the amalgamation property.*

Proof (1) follows from the equivalence between **Perfect** and the category of abelian ℓ-groups. To prove (2), suppose we have morphisms in **Perfect**, $f : A \to X \leftarrow B : g$. Let $\langle A, B\rangle = \{(a, b) \in A \times B \mid f(a) = g(b)\}$. It is easy to see that this set is a perfect MV-subalgebra of $A \times B$. Suppose for some perfect MV-algebra Y we have maps $\alpha : Y \to A$, $\beta : Y \to B$ such that $f\alpha = g\beta$. Define $h : Y \to \langle A, B\rangle\rangle$ by $h(y) = (\alpha(y), \beta(y))$ Then $\pi_1 h = f$, $\pi_2 h = g$. It follows that $\langle A, B\rangle$ is the pull-back of f along g. To prove (3), first observe that the direct product of two or more perfect MV-algebras need not be perfect, but it always contains perfect subalgebras. In particular there is, as a subalgebra of $\langle A, B\rangle\rangle$, its perfect skeleton Perf($\langle A, B\rangle\rangle$). It is straightforward to show that Perf($\langle A, B\rangle$) is indeed the product in the category **Perfect**. Statement (3) is then proved. To prove (4), let A, B', B'' be perfect MV-algebras and $\sigma' : A \hookrightarrow B'$, $\sigma'' : A \hookrightarrow B''$ embeddings. Then by [5] we know that the variety of all MV-algebras has the amalgamation property, so we have the following commutative diagram:

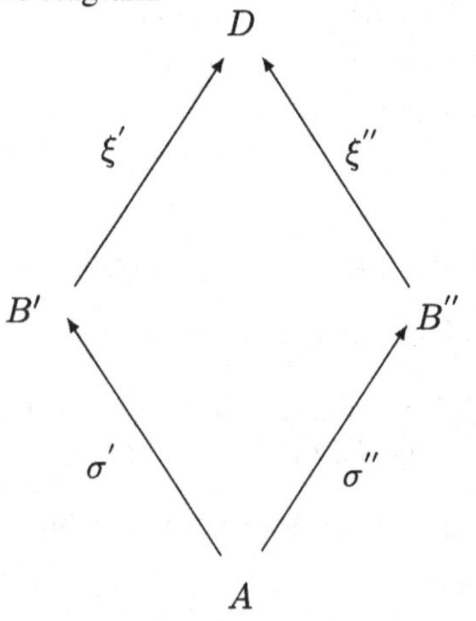

where ξ' and ξ'' are embeddings and D an MV-algebra. Since $\xi'(B')$ and $\xi''(B'')$ are perfect MV-algebras, then the following commutative diagram holds:

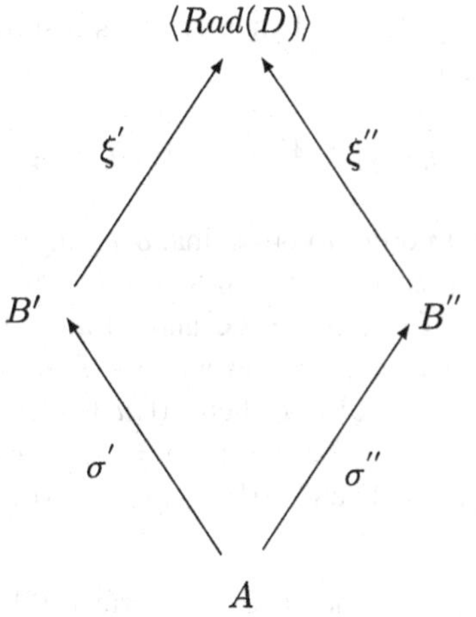

□

6.2 Ultraproduct of Perfect MV-Algebras

Let I an index set and $(A_i)_{i\in I}$ be a family of perfect MV-algebras. Let $A = \prod_{i\in I} A_i$ be the usual product in the category **MV** of MV-algebras. The category **Perfect** admits products too: the product of $(A_i)_{i\in I}$ in **Perfect** is the perfect skeleton of A, Perf(A). Set $A' = $ Perf(A). The elements of A' can be described as sequences $(a_i)'_{i\in I}$ such that $ord(a_i) = ord(a_j)$ for all $i, j \in I$.

Let F be a non principal ultrafilter in 2^I. In the category **MV** we have the usual ultraproduct A/F which consists of equivalence classes of sequences:

$$[(a_i)_{i\in I}] = [(b_i)_{i\in I}] \quad \text{iff} \quad \{i \mid a_i = b_i\} \in F.$$

Since perfect MV-algebras are first order definable (see below), we get that A/F is a perfect MV-algebra. On the other hand, we can consider the ultraproduct in the category **Perfect** by taking A'/F as the set of equivalence classes $[(a_i)'_{i\in I}]$ of elements $(a_i)'_{i\in I} \in A'$.

Proposition 6.16 *The algebras A/F and A'/F are isomorphic perfect MV-algebras.*

Let A be a perfect MV-algebra. A is called *locally archimedean* whenever $x, y \in Rad(A)$ and $nx \leq y$ for all positive integers n, then $x = 0$. A *weak unit* for A is a $w \in Rad(A)$ such that $w^{\perp} = \{0\}$. A will be called principal if $Rad(A)$ is a principal ideal.

Proof Since A' is a subalgebra of A then A'/F is a subalgebra of A/F. We can consider the inclusion map:

$$\sigma : [(a_i)'_{i\in I}] \in A'/F \hookrightarrow [(a_i)'_{i\in I}] \in A/F$$

that is a monomorphism. In order to prove that σ is surjective, let $[(a_i)_{i\in I}] \in A/F$. If $ord([(a_i)_{i\in I}]) = \infty$, then it is straightforward to prove that $\{i \mid ord(x_i) = \infty\} \in A/F$. Let $u_i = a_i$ if $ord(a_i) = \infty$ and let $u_i = 0$ if $ord(a_i) < \infty$. Then $(u_i)_{i\in I} \in A'$ since $ord(u_i) = \infty$ for every $i \in I$, so $[(u_i)_{i\in I}] \in A'/F$. Since $\{i \mid u_i = a_i\} = \{i \mid ord(a_i) = \infty\} \in F$ then $\sigma([(u_i)_{i\in I}]) = [(u_i)_{i\in I}] = [(a_i)_{i\in I}]$.

Similarly, if $ord([(a_i)_{i\in I}]) < \infty$ we let $u_i = a_i$ if $ord(a_i) < \infty$ and $u_i = 1$ otherwise. Again, $(u_i)_{i\in I} \in A'$ and $\sigma([(u_i)_{i\in I}]) = [(a_i)_{i\in I}]$. Therefore σ is an isomorphism. □

From the categorical equivalence between perfect MV-algebras and abelian ℓ-groups, given by the functors $\mathfrak{G}$ and $\mathfrak{D}$ it is reasonable to display the action of $\mathfrak{G}$ and $\mathfrak{D}$ focused on some special classes of perfect MV-algebras and abelian ℓ-groups. Indeed we consider the following subclasses of abelian ℓ-groups:

(1) the class of archimedean ℓ-groups (denoted by *Arch*);
(2) the class of archimedean ℓ-groups with a distinguished weak unit (denoted by $Arch_w$);
(3) the class of archimedean ℓ-groups with a distinguished strong unit (denoted by $Arch_s$).

The above classes of abelian ℓ-groups suggest to define classes of perfect MV-algebras reflecting, in the category of perfect MV-algebras, the role of the classes $Arch$, $Arch_w$ and $Arch_s$ in the category of abelian ℓ-groups.

Proposition 6.17 *Let A be a perfect locally archimedean MV-algebra, and let $a \in Rad(A)$. Then $A/(a^\perp)$ is locally archimedean.*

Proof Suppose $x, y \in Rad(A)$ and $nx \leq y(mod \quad a^\perp))$ for all n. Thus for all n, $((nx)\odot\neg y) \in a^\perp$. Therefore $(n(x\wedge a))\odot\neg y = 0$ for all n. But $(n(x\wedge a)\wedge na)\odot\neg y \leq (n(x\wedge a))\odot\neg y\wedge na = 0$, so we have $n(x\wedge a)\odot\neg y) = 0$ for all n. That is $n(x\wedge a) \leq y$ for all n. Since A is locally archimedean we have $x \wedge a = 0$. Thus $x \in a^\perp$, from which it follows that $A/(a^\perp)$ is locally archimedean. □

Lemma 6.18 *Let A be an MV-algebra and $w \in A$. Then $w^\perp = \{0\}$ iff for all $x, y \in A$, $x \wedge (y \oplus w) = y$ implies $x = y$.*

Proof Suppose for all $x, y \in A$, $x \wedge (y \oplus w) = y$ implies $x = y$. Let $x \in w^\perp$. Then, $x \wedge (0 \oplus w) = x \wedge w = 0$. Hence $x = 0$, so $w^\perp = \{0\}$. Conversely, let $w^\perp = \{0\}$. Suppose we have $x \wedge (y \oplus w) = y$. Then

$$0 = (\neg y) \odot y = (\neg y) \odot (x \wedge (y \oplus w)) = (\neg y) \odot x \wedge (\neg y) \odot (y \oplus w) = (\neg y) \odot x \wedge \neg y \wedge w.$$

Hence $(\neg y) \odot x \wedge \neg y = 0$. Now $(\neg y) \odot x \leq \neg y$, so $(\neg y) \odot x = 0$. Thus $x \leq yy$ and $x \leq y \oplus w$. So $x = x \wedge (y \oplus w) = y$. □

Corollary 6.19 *Let A be a perfect locally archimedean MV-algebra, and w a weak unit of A. Then $[(w, 0)] \in \mathfrak{D}(A)$ is also a weak unit.*

Proof Suppose $[(w, 0)] \wedge [(x, y)] = [(0, 0)]$. Then $x \wedge (y \oplus w) = y$, and by the above lemma, $x = y$. So, $[(x, y)] = [(0, 0)]$. □

Lemma 6.20 *Let A be a locally archimedean perfect MV-algebra. Then $\mathfrak{D}(A)$ is archimedean.*

Proof Let $[(x, y)] \in \mathfrak{D}(A)^+$ and assume $n[(x, y)] \leq [(a, b)]$ for all positive integers n. Then $[(a, b)]$ is positive, so we have, for all n, $n[(x \odot (\neg y), 0)] \leq [(a \odot (\neg b), 0)]$. Hence in A, $n(x \odot (\neg y)) \leq a \odot (\neg b) \in Rad(A)$, for all n. Thus $x \odot \neg y = 0$ and $x \leq y$. Now $[(x, y)]$ is positive in $\mathfrak{D}(A)$ iff $x \leq y$. So $x = y$ and $[(x, y)] = [(0, 0)]$. □

Lemma 6.21 *Let G be an abelian ℓ-group with a weak unit w. Then $(0, w)$ is a weak unit for $\mathfrak{G}(G)$.*

Proof Since $w \in G^+$, $(0, w) \in Rad(\mathfrak{G}(G))$. Therefore, for all $(0, x) \in Rad(\mathfrak{G}(G))$, if $(0, w) \wedge (0, x) = (0, w \wedge x) = 0$, then $x = 0$. So $(0, w)^{\perp} = \{0\}$. □

Lemma 6.22 *Let G be an archimedean ℓ-group. Then $\mathfrak{G}(G)$ is locally archimedean.*

Proof If $n(0, x) \leq (0, y)$ for all n, then $nx \leq y$, for all n, whence $x = 0$. □

Let **Arch** denote the full subcategory of **Perfect** whose objects are the locally archimedean perfect MV-algebras. From The above we get that **Arch** is equivalent to the full subcategory of abelian ℓ-groups, whose object are the archimedean ℓ-groups. Now let $\mathbf{Arch}_w$ be the category whose object are the pairs (A, w), where A is a locally archimedean perfect MV-algebra and w is a distinguished weak unit of A, and whose morphisms are the maps $f : (A, w) \to (A', w')$, where $f : A \to A'$ is an MV-homomorphism and $f(w) = w'$. Also we can define the category $\mathbf{AbArch}_w$ whose objects are archimedean ℓ-groups with a distinguished weak unit and whose morphisms are ℓ-groups homomorphisms preserving weak unit. Hence from above lemmas, corollaries and propositions we have:

Theorem 6.23 *The two correspondences*

$$(A, w) \mapsto (\mathfrak{D}(A), [(w, 0)] \quad \textit{and} \quad (G, u) \mapsto (\mathfrak{G}(G), (0, 0))$$

determine a categorical equivalence between $\mathbf{Arch}_w$ and $\mathbf{AbArch}_w$.

Also we can define the category $\mathbf{Arch}_s$ whose object are the pairs (A, p), where A is a locally archimedean perfect MV-algebra and $Rad(A) = id(p)$, and whose morphisms are the maps $f : (A, p) \to (A', p')$, where $f : A \to A'$ is an MV-homomorphism and $f(p) = p'$. In an analogous way, as the theorem above we get:

Theorem 6.24 *The two correspondences*

$$(A, p) \mapsto (\mathfrak{D}(A), [(p, 0)] \quad \textit{and} \quad (G, s) \mapsto (\mathfrak{G}(G), (0, s))$$

determine a categorical equivalence between $\mathbf{Arch}_w$ *and* $\mathbf{AbArch}_s$.

References

1. Chang, C.C.: Algebraic analysis of many-valued logics. Trans. Am. Math. Soc. **88**, 467–490 (1958)
2. Belluce, L.P., Di Nola, A., Lettieri, A.: Local MV-algebras, Rendiconti del Circolo Matematico di Palermo, Serie II. Tomo XLII, pp. 347–361 (1993)
3. Birkhoff, G.: Lattice Theory. Providence, Rhode Island (1967)
4. MacLane, S.: Categories for the Working Mathematicians. Springer, New York (1979)
5. Mundici, D.: The Haar theorem for lattice-ordered Abelian groups with order-unit. Discrete Contin. Dyn. Syst. **21**, 537–549 (2008)

Chapter 7
The Variety Generated by Perfect MV-Algebras

We remark that the functor Γ maps a non-equational class of groups, the category of abelian ℓ-groups with strong unit, to an equational class, the variety of all MV-algebras. On the other hand, the functor $\mathfrak{D}$ maps an equational class of groups, the category of abelian ℓ-groups, to a non-equational class, the category of Perfect MV-algebras. Also it is worth to remark that the class of perfect algebras does not form a variety, so the problem of studying the proper subvariety of the variety of all MV-algebras generated by all perfect MV-algebras arises.

Let $\mathcal{V}(\mathbb{Z})$ be the variety generated by the additive ℓ-group $\mathbb{Z}$ of integers with natural order, let $\mathcal{V}(\text{Perf})$ be the variety generated by all perfect algebras, and $\mathcal{V}(C)$ be the variety generated by Chang's algebra C. Then the following theorem holds:

Theorem 7.1 $\mathcal{V}(C) = \mathcal{V}(Perf)$

Proof For every perfect MV-algebra A, let $G = \mathfrak{D}(A)$ be its associated ℓ-group. Since the variety of abelian ℓ-groups is generated by $\mathbb{Z}$, then $G \in V(\mathbb{Z})$. Hence there exist an ℓ-homomorphism f and an abelian ℓ-group K such that $f(K) = G$ and $K \subseteq \mathbb{Z}^I$, for some set I, as an ℓ-group. From the equivalence between ℓ-groups and perfect MV-algebras $\mathfrak{G}(G) = \mathfrak{G}(f)(\mathfrak{G}(K))$ and $\mathfrak{G}(f)$ is an MV-homomorphism. Let the map $\rho : \mathfrak{G}(\mathbb{Z}^I) \hookrightarrow [\mathfrak{G}(\mathbb{Z})]^I$ be defined by $\rho(0, (z_i)_{i\in I}) = \{(0, z_i)\}_{i\in I}$ if $0 \le z_i$ for every $i \in I$ and $z_i \in \mathbf{Z}$; $\rho(1, (z_i)_{i\in I}) = \{(1, z_i)\}_{i\in I}$ if $z_i \le 0$ for every $i \in I$ and $z_i \in \mathbf{Z}$. Then ρ is an embedding. Since $\mathfrak{G}(\mathbb{Z}) \cong C$, then $\mathfrak{G}(K)$ is, up to isomorphism, a subalgebra of C^I and then $\mathfrak{G}(K) \in \mathcal{V}(C)$. Hence, $\mathfrak{G}(G)$ is a member of $\mathcal{V}(C)$ because it is obtained, by a homomorphism, from a member of $\mathcal{V}(C)$. Since $\mathfrak{G}(G) = \mathfrak{G}(\mathfrak{D}(A)) \cong A$, it follows that $A \in \mathcal{V}(C)$ because it is obtained as a homomorphic image of a member of $\mathcal{V}(C)$. Thus $\mathcal{V}(\text{Perf}) \subseteq \mathcal{V}(C)$. From $C \in \text{Perf}$ we get also that $\mathcal{V}(C) \subseteq \mathcal{V}(\text{Perf})$. The theorem is now proved. □

Theorem 7.2 *An MV-algebra A is in the variety $\mathcal{V}(C)$ iff A satisfies the identity:*

$$(x \oplus x) \odot (x \oplus x) = (x \odot x) \oplus (x \odot x). \qquad (*)$$

A. Di Nola et al., *Fuzzy Logic of Quasi-Truth: An Algebraic Treatment*,
Studies in Fuzziness and Soft Computing 338, DOI 10.1007/978-3-319-30406-9_7

Proof Let $\mathbf{K}$ denote the subvariety of MV-algebras defined by the identity $(*)$. Trivially $S_1^{\omega} = C \in \mathbf{K}$. Now we are going to prove that for $n \geq 2$, $S_n^{\omega} \notin \mathbf{K}$. Indeed, if n is even, then the element $\frac{1}{2} \in S_n$ does not satisfy the identity $(*)$. If n is odd, the identity $(*)$ fails in S_n^{ω} by the element $\frac{\frac{(n-1)}{2}}{n}$. As a consequence we get that $S_n^{\omega} \notin \mathbf{K}$ for every $n \geq 2$. By an application of ([1], Theorem 4.11) we get that $\mathbf{K} = \mathcal{V}(C) = \mathcal{V}(S_1^{\omega})$. □

Corollary 7.3 *Let A be a perfect non-Boolean MV-chain. Then $\mathcal{V}(A) = \mathcal{V}(Perf)$.*

Proof We know that $C \hookrightarrow A$, then $\mathcal{V}(C) \subseteq \mathcal{V}(A)$ But $A \in \mathcal{V}(C)$, the $\mathcal{V}(A) \subseteq \mathcal{V}(C) = \mathcal{V}(\text{Perf})$. □

Theorem 7.4 *Let $A \in \mathcal{V}(C)$. Then A is a subdirect product of perfect MV-chains.*

Proof By Chang's representation theorem A can be subdirectly embedded into $\prod_{P \in Spec(A)}(A/P)$. Let J be a prime ideal of A and M be a maximal ideal containing J. Then for every $x \in A$ we must have either $ord(x/P) = \infty$, or $ord((\neg x)/P) = \infty$. If $x \in M$ then $ord(x/P) = \infty$, otherwise there is $n \in \mathbb{N}$ such that $\neg(nx) \in P$ and then $(\neg x)^n \in M$, which is impossible, because from $x \in M$ and $(\neg x)^n \in M$ we get $\neg x \in M$, in contradiction with $x \in M$. If $x \notin M$, then $\neg x \in M$, and using the above argument, it is not hard to see that $ord((\neg x)/P) = \infty$. Therefore, for every $x/P \in A/P$ either $ord(x/P) = \infty$, or $ord((\neg x)/P) = \infty$. The fact that A/P is an MV-chain yields that A/P is perfect. □

Proposition 7.5 *Let A be an MV-algebra. Then the following are equivalent:*

(1) $A \in \mathcal{V}(C)$;
(2) For every maximal ideal M of A, $A = M \cup \neg(M)$.

Proof Let $A \in V(C)$. Then, for every $M \in Max(A)$, $\frac{A}{M} = \{0, 1\}$, that is for every $x \in A$ either $\frac{x}{M} = 0$ or $\frac{x}{M} = 1$, so $x \in M$ or $\neg x \in M$. Hence $A = M \cup \neg M$. Assume now that for every M, $A = M \cup \neg M$. We claim that A is a subdirect product of perfect MV-chains. Indeed, by Chang's representation theorem A can be subdirectly embedded into $\prod_{P \in Spec(A)} \frac{A}{P}$. Let $P \in Spec(A)$ and M the unique maximal ideal of A containing P. Then for every $x \in A$, either $ord(\frac{x}{P}) = \infty$ or $ord(\frac{\neg x}{P}) = \infty$. If $x \in M$, then $ord(\frac{x}{P}) = \infty$, otherwise there is $n \in N$ such that $\neg(nx) \in P$ and $(\neg x)^n \in M$. Hence from $x \in M$ and $(\neg x)^n \in M$ it follows $\neg x \in M$, which is absurd. If $\neg x \in M$, then $\neg x \in M$, and using the above argument, we get that $ord(\frac{\neg x}{P}) = \infty$. Therefore, for every $\frac{x}{P} \in \frac{A}{P}$ either $ord(\frac{x}{P}) = \infty$ or $ord(\frac{\neg x}{P}) = \infty$. Since $\frac{A}{P}$ is totally ordered, we get that $\frac{A}{P}$ is perfect. By Theorem 7.4, $A \in \mathcal{V}(C)$. □

Let H be a proper ideal of an MV-algebra A and A_H denote the subalgebra of A generated by H. Then $A_H = H \cup \neg H$, see [2]. Let $A_0 = \bigcap_{M \in Max(A)} A_M$. Then we have:

Proposition 7.6 *Let A be an MV-algebra and $B \in \mathcal{V}(C)$ a subalgebra of A, then*

(1) $A_0 \in \mathcal{V}(C)$;
(2) $B \subseteq A_0$.

Proof Let $y \in A_0$ and $M_0 \in Max(A_0)$. Then $y \in A_M$ for every $M \in Max(A)$. Let $N \in Max(A)$ be such that $M_0 = N \cap A_0$. Then $y \in N \cup \neg N$ and hence $y \in (N \cup \neg N) \cap A_0$, i.e., $y \in M_0 \cup (\neg N \cap A_0) = M_0 \cup \neg M_0$. Hence $y \in A_{0M_0}$ and then $A_0 \subseteq A_{0M_0}$, i.e., $A_0 = A_{0M_0}$ for every $M_0 \in Max(A_0)$. This yelds (1). To prove (2), let $M \in Max(A)$ and $N = B \cap M$. Then N is a maximal ideal of B. From $B \in V(C)$ we get that $B = N \cup \neg N$ and $B = (M \cap B) \cup (\neg M \cap B) = A_M \cap B$. Thus, $B \subseteq A_M$ for every $M \in Max(A)$. Hence $B \subseteq \bigcap_{M \in Max(A)} A_M = A_0$. □

We call A_0 the $\mathcal{V}(C)$-*skeleton* of A. An ideal I of an MV-algebra A is called $\mathcal{V}(C)$-*ideal* if and only if for every maximal ideal M of A, $I \subseteq M$ implies $A = M \cup \neg(M)$.

Theorem 7.7 *Let A an MV-algebra and I and ideal of A. The following are equivalent:*

(1) $A/I \in \mathcal{V}(C)$;
(2) I *is a* $\mathcal{V}(C)$-ideal *of* A.

Proof (1) implies (2). Let $A/I \in \mathcal{V}(C)$ and $M \in Max(A)$ such that $I \subseteq M$. Then M/I is maximal ideal of A/I. By hypothesis, $A/I = (M/I) \cup (\neg(M))/I)$. Let $x \in A$, then either $x/I \in M/I$ or $x/I \in (\neg M)/I$. Hence, we must either have $x \in M$ or $x \in neg(M)$, whence $A = M \cup \neg(M)$.

(2) implies (1). Let M/I be a maximal ideal of A/I. Then $I \subseteq M$, and $M \in Max(A)$. By hypothesis, $A = M \cup \neg(M)$, then $A/I = (M/I) \cup (\neg(M))/I)$ for every maximal ideal M/I of A/I. Therefore $A/I \in \mathcal{V}(C)$. □

Corollary 7.8 *Let A be an MV-algebra. Then the following are equivalent:*

(1) each ideal of A is a $\mathcal{V}(C)$-ideal;
(2) $A \in \mathcal{V}(C)$.

Proof Trivial. □

Theorem 7.9 *Let A be an MV-algebra. Then its $\mathcal{V}(C)$-skeleton, A_0, is generated by the subset $B(A) \cup Rad(A)$ of A.*

Proof Since $Rad(A) \subseteq A_0$ and $B(A) \subseteq A_0$, then $\langle B(A) \cup Rad(A) \rangle \subseteq A_0$. A_0 can subdirectly be embedded into a direct product $\prod_{i \in I} A_i$, where A_i is a perfect MV-chain, for each $i \in I$. So, every $x \in A$ can be written as $x = (x_i)_{i \in I}, x_i \in A_i$. Let 0_i and 1_i denote the first and the last element of A_i, respectively. For every $x \in A_0$ it can be easily checked that:

(i) $2(x \odot x) \in B(A)$,
(ii) $x \wedge \neg x \in Rad(A)$,
(iii) $x = (2(x \odot x)) \odot (x \wedge \neg x) \oplus (x \odot x) \odot (x \vee \neg x)$.

If $x_i \in Rad(A_i)$, then by (i) and (ii) we get $x_i = ((2(x \odot x)) \odot (x \wedge \neg x) \oplus (x \odot x) \odot (x \vee \neg x))_i$. In an analogous way it can be seen that if $x_i \in \neg Rad(A_i)$ it is $x_i = ((2(x \odot x)) \odot (x \wedge \neg x) \oplus (x \odot x) \odot (x \vee \neg x))_i$. Thus by (i), (ii) and (iii) it follows that $x \in \langle B(A) \cup Rad(A) \rangle$, and then $A_0 \subseteq \langle B(A) \cup Rad(A) \rangle$. □

7.1 Quasi Variety Generated by C

In this section, we show that the quasi variety generated by Chang algebra C coincides with the variety generated by C.

Theorem 7.10 $\mathcal{V}(C) = \mathcal{QV}(C)$.

To prove the previous theorem, we give some auxiliary results.

Lemma 7.11 $\Gamma(Z \times_{lex} Q, (1, 0)) \in \mathcal{QV}(C)$.

Proof Let us suppose that $A = \Gamma(Z \times_{lex} Q, (1, 0))$. Suppose a quasi-identity $p(x) = 0 \to q(x) = 0$ is *false* in A. We suppose p, q are polynomials in one variable (the case of n variables is analogous). Then there is x such that $p(x) = 0$ and $q(x) \neq 0$. We can suppose $x \in Rad(A)$ and $x \neq 0$. But then x generates a copy of C. So, the quasi-identity is false also in C. □

Corollary 7.12 $\Gamma(Z \times_{lex} R, (1, 0)) \in \mathcal{QV}(C)$.

Proof This follows by the density of the rationals in R. □

Corollary 7.13 *If* $*R$ *is an ultrapower of the reals, then*

$$\Gamma(Z \times_{lex} *R, (1, 0)) \in \mathcal{QV}(C).$$

Proof This follows from Los Theorem (Proposition 2.7) on ultraproducts. □

Corollary 7.14 *If* G *is any linearly ordered abelian group, then*

$$\Gamma(Z \times_{lex} G, (1, 0)) \in \mathcal{QV}(C).$$

Proof This follows because every linearly ordered abelian group embeds in an ultrapower of the reals. □

Corollary 7.15 *Every perfect* MV *chain is in* $\mathcal{QV}(C)$.

Proof This follows because every perfect MV-chain has the form $\Gamma(Z \times_{lex} G, (1, 0))$. □

Now, let us return to the proof of Theorem 7.10. Clearly $\mathcal{QV}(C) \subseteq \mathcal{V}(C)$. Conversely, an MV-chain belongs to $\mathcal{V}(C)$ if and only if it is perfect, so every MV-chain belonging to $\mathcal{V}(C)$ belongs to $\mathcal{QV}(C)$. But every element of $\mathcal{V}(C)$ is a subdirect product of chains of $\mathcal{V}(C)$, and $\mathcal{QV}(C)$ is closed under subdirect products. So, $\mathcal{V}(C) \subseteq \mathcal{QV}(C)$. Hence $\mathcal{V}(C) = \mathcal{QV}(C)$, and the proof is complete.

References

1. Komori, Y.: Super-Łukasiewicz propositional logic. Nagoya Math. J. **84**, 119–133 (1981)
2. Hoo, C.S.: MV-algebras, ideals and semisimplicity. Math. Japon. **34**(4), 563–583 (1989)

Chapter 8
Representations of Perfect MV-Algebras

8.1 Gödel Spaces

In the sequel we denote by $\mathbf{MV(C)}$ the category of the class of objects coincides with the variety $\mathcal{V}(C)$, generated by C, and morphisms are algebraic homomorphisms, and the variety $\mathcal{V}(C)$. We extract from the variety $\mathbf{MV(C)}$ the subclass $\mathbf{MV(C)^G}$ generated by $MV(C)$-algebras C_n, $0 \leq n < \omega$, by means of the operators of direct products, subalgebras and direct limits. The category of Gödel spaces $\mathcal{GS}$ (with strongly isotone maps as morphisms), which are dually equivalent to the category of Gödel algebras, is transferred by a contravariant functor $\mathcal{H}$ into the category $\mathbf{MV(C)^G}$. Conversely, the category $\mathbf{MV(C)^G}$ is transferred into the category $\mathcal{GS}$ by means of a contravariant functor $\mathcal{P}$. Moreover, it is shown that the functor $\mathcal{H}$ is faithful, the functor $\mathcal{P}$ is full and the both functors are dense. The description of finite coproduct of algebras, which are isomorphic to Chang algebra, is given. Using duality a characterization of projective algebras in $\mathbf{MV(C)^G}$ is given.

Recall some notations: let $C_0 = \Gamma(Z, 1)$, $C_1 = C \cong \Gamma(Z \times_{lex} Z, (1, 0))$ with generator $(0, 1) = c_1 (= c)$, $C_m = \Gamma(Z \times_{lex} \cdots \times_{lex} Z, (1, 0, \ldots, 0))$ with generators $c_1 (= (0, 0, \ldots, 1)), \ldots, c_m (= (0, 1, \ldots, 0))$, where the number of factors Z is equal to $m \geq 1$ and $\times_{lex}$ is the lexicographic product. Let us denote $Rad(A) \cup \neg Rad(A)$ through $R^*(A)$.

We are interested in the class $\mathbf{LSP}\{C_i : i \in \omega\}$ of $MV(C)$-algebras which is generated by the set $\{C_i : i \in \omega\}$ by the operators of direct products, subalgebras and direct limits, where C_0 is two-element Boolean algebra, $C_1 = C$ and C_n $(n > 1)$ is n-generated perfect MV-chain.

Let $\mathbf{K}$ be any variety of algebras. Then $F_{\mathbf{K}}(m)$ denotes the m-generated free algebra in the variety $\mathbf{K}$.

Now we introduce the notion of *weak duality* between categories. Let $\mathbf{A}, \mathbf{B}$ be categories. We say that $\mathbf{A}$ and $\mathbf{B}$ are *weakly dual* (or that there is a *weak duality* between $\mathbf{A}$ and $\mathbf{B}$) if there are dense contravariant functors $F_{\mathbf{A}} : \mathbf{A} \to \mathbf{B}$ and $F_{\mathbf{B}} : \mathbf{B} \to \mathbf{A}$ such that $F_{\mathbf{A}}$ is faithful and $F_{\mathbf{B}}$ is full.

A. Di Nola et al., *Fuzzy Logic of Quasi-Truth: An Algebraic Treatment*,
Studies in Fuzziness and Soft Computing 338, DOI 10.1007/978-3-319-30406-9_8

In this section, we give the description of m-generated free algebras in the variety $\mathbf{MV(C)}$ generated by perfect MV-algebras. We describe the category of Gödel spaces, where any Gödel space is a special case of Priestley spaces. We also will prove that there is a weak duality between the full subcategory $\mathbf{MV(C)}^{\mathbf{G}}(= \mathbf{LSP}\{C_i : i \in \omega\})$ of the category $\mathbf{MV(C)}$ and the category of Gödel spaces $\mathcal{GS}$. More precisely, we construct the functors $\mathcal{P} : \mathbf{MV(C)}^{\mathbf{G}} \to \mathcal{GS}$, which is full, and $\mathcal{H} : \mathcal{GS} \to \mathbf{MV(C)}^{\mathbf{G}}$ which is faithful.

In the category theory, a functor $\mathcal{F} : \mathbf{E} \to \mathbf{D}$ is *dense* (or *essentially surjective*) if each object D of $\mathbf{D}$ is isomorphic to an object of the form $\mathcal{F}(E)$ for some object E of $\mathbf{E}$. The suggested functors $\mathcal{P} : \mathbf{MV(C)}^{\mathbf{G}} \to \mathcal{GS}$ and $\mathcal{H} : \mathcal{GS} \to \mathbf{MV(C)}^{\mathbf{G}}$ are dense.

The category $\mathcal{GS}$ of Gödel spaces is dually equivalent to the category $\mathbf{GA}$ of Gödel algebras. Hence, there exist two functors $\mathcal{G} : \mathbf{GA} \to \mathcal{GS}$ and $\mathcal{HS} : \mathcal{GS} \to \mathbf{GA}$. So, we also have two functors $\mathcal{HS} \circ \mathcal{P} : \mathbf{MV(C)}^{\mathbf{G}} \to \mathbf{GA}$ and $\mathcal{H} \circ \mathcal{G} : \mathbf{GA} \to \mathbf{MV(C)}^{\mathbf{G}}$. Moreover, $\mathcal{HS} \circ \mathcal{P}$ coincides with Belluce functor β [1] defined on the $\mathbf{MV(C)}^{\mathbf{G}}$.

Using the weak duality we give a construction of a coproduct in the variety $\mathbf{MV(C)}^{\mathbf{G}}$ which coincides with coproduct in $\mathbf{MV(C)}$. Moreover, we show that the coproduct coincides with free product (using this weak duality). Free products in various classes of ℓ-groups were investigated in the frame of varieties of ℓ-groups or abelian ℓ-groups by Holland and Scrimger [2], Martinez [3, 4], Powel and Tsinakis [5], Mundici [6], Dvurecenskij and Holland [7], Di Nola and Lettieri [8]. Moreover, D. Mundici in [6] has shown that coproduct coincides with free product in the variety of MV-algebras.

We notice that in [9] it is established a duality between the category of finitely generated $MV(C)$-algebras, having finite spectrum, and the category of finite dual Heyting algebras which satisfy linearity condition.

A *Boolean space* is zero-dimensional, compact and Hausdorff topological space. The category of Boolean spaces and continuous maps is denoted by $\mathcal{B}$. A Priestley space is a triple $(X; R, \Omega)$, where $(X; \Omega)$ is a Boolean space and R is an order relation on X such that, for all $x, y \in X$ with $x\bar{R}y$, there exists a clopen up-set V with $x \in V$ and $y \notin V$. A morphism between Priestley spaces is a continuous order-preserving map. We denote the category of Priestley spaces plus continuous order-preserving maps by $\mathcal{PS}$. For details on Priestley duality see Priestley [10] and Davey and Priestley [11]. Note that for simplicity sake we will often refer to a Boolean or Priestley space by its underlying set X.

Priestley duality relates the category of bounded distributive lattices to the category of Priestley spaces by mapping each bounded distributive lattice L to its ordered space $\mathcal{F}(L)$ of prime filters, and mapping each Priestley space X to the bounded distributive lattice $\mathcal{P}(X)$ of clopen up-sets of X. When restricted to Heyting algebras and Heyting spaces respectively, these mappings give the restricted Priestley duality for Heyting algebras.

A Heyting algebra is an algebra $(A, \vee, \wedge, \to, 0, 1)$ of type $(2, 2, 2, 0, 0)$, where $(A, \vee, \wedge, 0, 1)$ is a bounded distributive lattice and the binary operation $\to$, which is called implication, satisfies

$$(\forall a, b, x \in A)(x \wedge a \leq b \Leftrightarrow x \leq a \rightarrow b).$$

The following facts are easily proved: (a) every finite distributive lattice is (the underlying lattice of) a Heyting algebra, (b) every distributive algebraic lattice is a Heyting algebra, (c) the lattice of all open subsets of a topological space forms a Heyting algebra.

A Heyting space (or Esakia space, in other terminology) X is a Priestley space such that $R^{-1}(U)$ is open for every open subset U of X. (Recall that $R^{-1}(U) = \{y \in X : (\exists u \in U) yRu\}$ and that $R^{-1}\{x\}$ is abbreviated to $R^{-1}(x)$). The sets $R(U)$ and $R(x)$ are defined dually.) A morphism between Heyting spaces, called a *strongly isotone map* (or *Heyting morphism* in other terminology), is a continuous map $\varphi : X \rightarrow Y$ such that $\varphi(R(x)) = R(\varphi(x))$ for all $x \in X$. The restricted Priestley duality for Heyting algebras states that a bounded distributive lattice A is the underlying lattice of a Heyting algebra if and only if the Priestley dual of A is a Heyting space, and that a $\{0, 1\}$-lattice homomorphism h between Heyting algebras preserves the operation $\rightarrow$ if and only if the Priestley dual of h is a Heyting morphism. We denote the category of Heyting spaces plus Heyting morphisms by $\mathcal{HS}$.

For any Priestley space (X, R) we define $\mathcal{P}(X)$ as the set of all clopen up-sets of X. For any $U, V \in \mathcal{P}(X)$ define: $U \vee V = U \cup V$ and $U \wedge V = U \cap V$. Then the algebra $\mathcal{P}((X, R)) = (\mathcal{P}(X), \vee, \wedge, \emptyset, X)$ is a bounded distributive lattice. Furthermore, for any morphism $f : (X_1, R_1) \rightarrow (X_2, R_2)$ in $\mathcal{PS}$, $\mathcal{F}(f) = f^{-1}$ is a $\{0, 1\}$-lattice homomorphism from $\mathcal{P}((X_2, R_2))$ into $\mathcal{P}(X_1, R_1)$. On the other hand, for each bounded distributive lattice L, the set $\mathcal{F}(L)$ of all prime filters of L with the binary relation R on it,which is the inclusion between prime filters, and topologised by taking the family of $supp^*(a) = \{F \in \mathcal{F}(L) : a \in F\}$, for $a \in L$, and their complements as a subbase, is an object of $\mathcal{PS}$; and for each $\{0, 1\}$-lattice homomorphism $h : L_1 \rightarrow L_2$, $\mathcal{F}(h) = h^{-1}$ is a morphism of $\mathcal{PS}$. Therefore, we have two contravariant functors $\mathcal{F} : \mathbf{D} \rightarrow \mathcal{PS}$ and $\mathcal{P} : \mathcal{PS} \rightarrow \mathbf{D}$. These functors establish a dual equivalence between the categories of bounded distributive lattices $\mathbf{D}$ and Priestley spaces $\mathcal{PS}$.

For any Heyting space (X, R) and $U, V \in \mathcal{H}(X)$(= the set of all clopen up-sets of X) define:

$$U \rightarrow V = X \backslash (R^{-1}(U \backslash V))$$

Then the algebra $\mathcal{H}((X, R)) = (\mathcal{H}(X), \vee, \wedge, \rightarrow, \emptyset, X)$ is a Heyting algebra. Furthermore, for any morphism $f : (X_1, R_1) \rightarrow (X_2, R_2)$ in $\mathcal{HS}$, $\mathcal{H}(f) = f^{-1}$ is a Heyting algebra homomorphism from $\mathcal{H}((X_2, R_2))$ into $\mathcal{H}(X_1, R_1)$. On the other hand, for each Heyting algebra A, the set $\mathcal{F}(A)$ of all prime filters of A with the binary relation R on it,which is the inclusion between prime filters, and topologized by taking the family of $supp^*(a) = \{F \in \mathcal{F}(A) : a \in F\}$, for $a \in A$, and their complements as a subbase, is an object of $\mathcal{HS}$; and for each Heyting algebra homomorphism $h : A \rightarrow B$, $\mathcal{F}(h) = h^{-1}$ is a morphism of $\mathcal{HS}$. Therefore, we have two contravariant functors $\mathcal{F} : \mathbf{HA} \to \mathcal{HS}$ and $\mathcal{H} : \mathcal{HS} \to \mathbf{HA}$. These functors establish a dual equivalence between the categories **HA** and $\mathcal{HS}$.

A Heyting algebra A is said to be *Gödel algebra* (or $\mathcal{L}$-algebra [12]) if it satisfies the linearity condition: $(a \to b) \vee (b \to a) = 1$ for all $a, b \in A$. Gödel algebras represent the algebraic models for Gödel logic G. It is well known that the Heyting spaces for Gödel algebras form root systems. A. Horn [13] showed that Gödel algebras can be characterized among Heyting algebras in terms of the order on prime filters (co-ideals). Specifically, a Heyting algebra is a Gödel algebra iff its set of prime lattice filters is a root system (ordered by inclusion). So we can define a *Gödel space* X as a Heyting space such that $R(x)$ is a chain for any $x \in X$. The category of Gödel spaces and strongly isotone maps is denoted by $\mathcal{GS}$. Also, we denote the category of Gödel algebras by **GA**.

8.2 M-Generated Free $MV(C)$-Algebra

Recall that an MV-algebra $A = (A, 0, \neg, \oplus)$ is an abelian monoid $(A, 0, \oplus)$ equipped with a unary operation $\neg$ such that $\neg\neg x = x$, $x \oplus \neg 0 = \neg 0$, and $y \oplus \neg(y \oplus \neg x) = x \oplus \neg(x \oplus y)$ [14]. We set $1 = \neg 0$ and $x \odot y = \neg(\neg x \oplus \neg y)$ [15]. We shall write ab for $a \odot b$ and a^n for $\underbrace{a \odot \cdots \odot a}_{n \text{ times}}$, for given $a, b \in A$. Every MV-algebra has an underlying ordered structure defined by

$$x \leq y \text{ iff } \neg x \oplus y = 1.$$

Then $(A; \leq, 0, 1)$ is a bounded distributive lattice. Moreover, the following property holds in any MV-algebra:

$$xy \leq x \wedge y \leq x \vee y \leq x \oplus y.$$

The unit interval of real numbers $[0, 1]$ endowed with the following operations: $x \oplus y = \min(1, x + y)$, $x \odot y = \max(0, x + y - 1)$, $\neg x = 1 - x$, becomes an MV-algebra. It is well known that the variety **MV** of all MV-algebras is generated by the MV-algebra $S = ([0, 1], \oplus, \odot, \neg, 0, 1)$, i.e. $\mathcal{V}(S) = \mathbf{MV}$.

The algebra C, with generator $c \in C$, is isomorphic to $\Gamma(Z \times_{lex} Z, (1, 0))$, with generator $(0, 1)$. Recall also that the intersection of all maximal ideals of an MV-algebra A, the radical of A, is denoted by $Rad(A)$.

Theorem 8.1 *An* 1*-generated free* $MV(C)$*-algebra* $F_{\mathbf{MV(C)}}(1)$ *is isomorphic to* C^2 *with free generator* $(c, \neg c)$.

Proof Firstly, let us show that C^2 is generated by $(c, \neg c)$. Indeed, $2((c, \neg c)^2) = (0, 1)$ and $(2(c, \neg c))^2 = (1, 0)$. Therefore, since c (and $\neg c$, as well) generates C, we have that $(c, \neg c)$ generates C^2.

Observe that if we have a perfect $MV(C)$-chain A, then 1-generated subalgebra of A is isomorphic to either $\Gamma(Z \times_{lex} Z, (1, 0))$ or the two-element Boolean algebra S_1.

Let **K** be a variety. An m-generated free algebra A on the generators $g_1, \ldots, g_m$ over the variety **K** can be defined in the following way: the algebra A is a free m-generated algebra on the generators $g_1, \ldots, g_m$ iff for any m-variable equation $P(x_1, \ldots, x_m) = Q(x_1, ..., x_m)$, the equation holds in the variety **K** iff the equation $P(g_1, \ldots, g_m) = Q(g_1, \ldots, g_m)$ is true in the algebra A (on the generators $g_1, \ldots, g_m \in A$) [16].

Now, suppose that one-variable equation $P = Q$ does not hold in the variety **MV(C)**. It means that this equation does not hold in some 1-generated perfect $MV(C)$-algebra A on some element $a \in A$. Then A is isomorphic either to C or S_1 (2-element Boolean algebra). Let us suppose that A is isomorphic to C. Identify isomorphic elements. Depending on the generator of A, the one belongs to either $Rad A$ or $\neg Rad A$, we use the projection either $\pi_1 : C^2 \to C$ or $\pi_2 : C^2 \to C$, sending the generator $(c, \neg c)$ either to $c \in C$ or to $\neg c \in C$. From here we conclude that $P = Q$ does not hold in C^2. Now let us suppose that A is isomorphic to S_1. Notice that homomorphic image of C^2 by $Rad(C^2)$ is isomorphic to one-generated free Boolean algebra S_1^2. So, $P = Q$ does not hold in C^2. Hence, C^2 is 1-generated free $MV(C)$-algebra. □

As we know the algebra C_n is generated by n generators $c_1, c_2, \ldots, c_n \in Rad(C_n)$. In general, C_n is generated by n generators $c_{\varphi_i(1)}, c_{\varphi_i(2)}, \ldots, c_{\varphi_i(n)}$ for any $i \in \{1, \ldots, n!\}$, where $\varphi_i : \{1, \ldots, n\} \to \{1, \ldots, n\}$ is any bijection: the first generator is $c_{\varphi_i(1)}$, the second generator is $c_{\varphi_i(2)}$ and so on, the n-th generator is $c_{\varphi_i(n)}$. Denote $(c_{\varphi_i(1)}, c_{\varphi_i(2)}, \ldots, c_{\varphi_i(n)})$ by $\mathbf{a_i}$. As we see we have $n!$ different sets of ordered generators that generate C_n. Now let us consider the algebra $C_n^{n!}$ and the subalgebra B_n of the algebra $C_n^{n!}$ generated by n generators $\mathbf{b_i} = (\pi_i(\mathbf{a_1}), \pi_i(\mathbf{a_2}), \ldots, \pi_i(\mathbf{a_{n!}}))$, $i = 1, \ldots, n$. Notice that the generators $\mathbf{b_1}, \ldots, \mathbf{b_n}$ belong to $Rad(C_n^{n!})$. Therefore, the algebra B_n is perfect. Moreover, any j-th factor ($j \in \{1, \ldots, n!\}$) $\pi_j|_{B_n}$ is isomorphic to C_n, since $\pi_j|_{B_n}(\mathbf{b_1})(= \pi_1(\mathbf{a_j})), \pi_j|_{B_n}(\mathbf{b_2})(= \pi_2(\mathbf{a_j})), \ldots, \pi_j|_{B_n}(\mathbf{b_n})(= \pi_n(\mathbf{a_j}))$ generate $\pi_j|_{B_n}(B_n)(\cong C_n)$, where $\pi_j|_{B_n}$ is the restriction of the projection $\pi_j : C_n^{n!} \to C_j$ $(j = 1, \ldots, n!)$ on the subalgebra B_n.

Let us consider the subalgebra A_k of the algebra $\Pi_{i=1}^{\infty} D_i^{(k)}$, where $D_i^{(k)} \cong C_k$ $(1 \leq k < n)$, generated by $\mathbf{d_j^{(k)}} = (u_{1k}^{(j)}, u_{2k}^{(j)}, u_{3k}^{(j)}, ..., u_{ik}^{(j)}, ...)$, $j = 1, \ldots, n$ where $u_{ik}^{(1)}, \ldots, u_{ik}^{(n)} \in Rad(D_i^{(k)})$ generate $D_i^{(k)}$, $(u_{ik}^{(1)}, \ldots, u_{ik}^{(n)}) \neq (u_{jk}^{(1)}, \ldots, u_{jk}^{(n)})$ for $i \neq j$.

Let $B(n)$ be a subalgebra of $B_n \times A_1 \times \cdots \times A_{n-1}$ generated by

$$\mathbf{g_1} = (\mathbf{b_1}, \mathbf{d_1^{(1)}}, \ldots, \mathbf{d_1^{(n-1)}}), \ldots, \mathbf{g_n} = (\mathbf{b_n}, \mathbf{d_n^{(1)}}, \ldots, \mathbf{d_n^{(n-1)}}).$$

Notice that the generators $\mathbf{g_1}, \ldots, \mathbf{g_n}$ belong to $Rad(B_n \times A_1 \times \cdots \times A_{n-1})$. Therefore, the algebra $B(n)$ is perfect. Observe that $B(n)$ is also generated by $\mathbf{g}_1^{\varepsilon_{i1}}, \ldots, \mathbf{g}_n^{\varepsilon_{in}}$, where $\varepsilon_{i1}, \ldots, \varepsilon_{in}$ is any sequence of 1 and 0, $1 \leq i \leq 2^n$, and $x^{\varepsilon} = \begin{cases} x, & if \ \varepsilon = 1 \\ \neg x, & if \ \varepsilon = 0 \end{cases}$. Hence we have

Lemma 8.2 *The algebra $B(n)^{2^n}$ is generated by* $G_1 = (\mathbf{g}_1^{\varepsilon_{11}}, \mathbf{g}_1^{\varepsilon_{21}}, \dots, \mathbf{g}_1^{\varepsilon_{2^n 1}})$, $G_2 = (\mathbf{g}_2^{\varepsilon_{12}}, \mathbf{g}_2^{\varepsilon_{22}}, \dots, \mathbf{g}_2^{\varepsilon_{2^n 2}}), \dots, G_n = (\mathbf{g}_n^{\varepsilon_{1n}}, \mathbf{g}_n^{\varepsilon_{2n}}, \dots, \mathbf{g}_n^{\varepsilon_{2^n n}})$.

Proof Observe that $\mathbf{g}_i \in Rad(B(n))$ and $\neg\mathbf{g}_i \in \neg Rad(B(n))$. Therefore $(2\mathbf{g}_i)^2 = 0$ and $(2\neg\mathbf{g}_i)^2 = 1$. So, $(2G_i)^2$ is a 2^n element sequence of 0 and 1 which represents free generators of n-generated free Boolean algebra $\mathbf{2}^{2^n}$. By means of this free Boolean generators, we obtain all 2^n-element sequence of 0 and 1. Taking into account that any i-th factor of $B(n)^{2^n}$ is generated by $\pi_i(G_1), \dots, \pi_i(G_n)$, we conclude that $B(n)^{2^n}$ is generated by $G_1, \dots, G_n$. □

Observe, that, according to the construction of the algebra $B(n)$, if the chain $MV(C)$-algebra A is generated by n generators from $Rad(B(n))$, then A is a homomorphic image of $B(n)$ sending the generators of $B(n)$ to the generators of A, since $B(n)$ contains as a factor all such kind of chains. So, we have

Lemma 8.3 *If a chain $MV(C)$-algebra A is generated by n generators, then A is a homomorphic image of $B(n)^{2^n}$, sending the generators of $B(n)$ to the generators of A.*

Proof Let us suppose that A is n-generated chain $MV(C)$-algebra. Then A coincides with some $D_i^{(k)} (\cong C_k)$, $1 \leq k < n$, generated by some $u_{ik}^{(1)}, \dots, u_{ik}^{(n)}$. But $D_i^{(k)}$ is a homomorphic image of $B(n)^{2^n}$. □

Theorem 8.4 *The n-generated free $MV(C)$-algebra $F_{\mathbf{MV(C)}}(n)$ is isomorphic to B^{2^n} with free generators*

$$G_1 = (\mathbf{g}_1^{\varepsilon_{11}}, \mathbf{g}_1^{\varepsilon_{21}}, \dots, \mathbf{g}_1^{\varepsilon_{2^n 1}}),$$
$$G_2 = (\mathbf{g}_2^{\varepsilon_{12}}, \mathbf{g}_2^{\varepsilon_{22}}, \dots, \mathbf{g}_2^{\varepsilon_{2^n 2}}),$$
$$\dots$$
$$G_n = (\mathbf{g}_n^{\varepsilon_{1n}}, \mathbf{g}_n^{\varepsilon_{2n}}, \dots, \mathbf{g}_n^{\varepsilon_{2^n n}}).$$

Proof We should prove that any n-variable equation $P(x_1, \dots, x_n) = Q(x_1, \dots, x_n)$ holds in the variety $\mathbf{MV(C)}$ if and only if $P(G_1, \dots, G_n) = Q(G_1, \dots, G_n)$ is true in the algebra $B(n)^{2^n}$. It is obvious that if n-variable equation $P(x_1, \dots, x_n) = Q(x_1, \dots, x_n)$ holds in the variety $\mathbf{MV(C)}$, then $P(G_1, \dots, G_n) = Q(G_1, \dots, G_n)$ is true in the algebra $B(n)^{2^n}$.

Now let us suppose that n-variable equation $P(x_1, \dots, x_n) = Q(x_1, \dots, x_n)$ does not hold in the variety $\mathbf{MV(C)}$. It means that this equation does not hold in some n-generated chain perfect $MV(C)$-algebra D on some element $d_1, \dots, d_n \in D$. Then D is isomorphic to either $S_1, C_1, \dots, C_{n-1}$ or C_n, where S_1 is two-element Boolean algebra. Identify the isomorphic elements. According to Lemma 8.3, there exists a homomorphism onto $f : B(n)^{2^n} \to D$ from $B(n)^{2^n}$ onto D such that $f(G_i) = d_i$. Since $P(d_1, \dots, d_n) \neq Q(d_1, ..., d_n)$ in D, we have that $P(G_1, \dots, G_n) \neq Q(G_1, \dots, G_n)$ in $B(n)^{2^n}$. From here we conclude that $B(n)^{2^n}$ is n-generated free $MV(C)$-algebra with free generators $G_1, \dots, G_n$. □

Recall that an algebra A is subdirectly irreducible iff A is trivial or there is the only atom in the lattice of all congruences $ConA$. In this case the least element is $\bigcap(ConA - \{\Delta\})$, a principal congruence [17], where Δ is the least element in the lattice $ConA$.

Lemma 8.5 *If a totally ordered $MV(C)$-algebra A is finitely generated, then A is subdirectly irreducible.*

Proof Let A be a totally ordered n-generated $MV(C)$-algebra. It means that there exist different elements $b_1, b_2, \ldots, b_n \in A$ such that the elements generate A and $b_i \neq 1$, $2b_i = 1$ for every $i \in \{1, \ldots, n\}$. Let us suppose that $b_1 > b_2 > \cdots > b_n$. Then there exists the sequence of proper principal filter $[b_1) \subseteq [b_2) \subseteq \cdots \subseteq [b_n)$ such that $[b_i) \neq \{1\}$ for every $i \in \{1, \ldots, n\}$. Therefore $\bigcap_{i=1}^{n}[b_i) = [b_1)$ that means that A is subdirectly irreducible. □

The inverse of the Lemma 8.5 is not true. Indeed, let us consider the direct limit C_ω of the direct system $\{C_i : i \in \omega, \varepsilon_{ij}, i \leq j\}$, where $C_i \ni c_k \mapsto \varepsilon_{ij}(c_k) = c_k \in C_j$ for $k \leq i$. It is obvious that C_ω is not finitely generated. Identifying isomorphic elements we have that C_ω is generated by $c_1, c_2, c_3, \ldots$. Nevertheless C_ω is subdirectly irreducible since $\bigcap_{i\in\omega}[\neg c_i) = [\neg c_1)$.

8.3 Spectral Duality

It is well known that the category **D** of bounded distributive lattices and bounded lattice homomorphisms, and the category **Spec** of spectral spaces and spectral maps (strongly continuous maps) are dually equivalent. Since **D** is dually equivalent to both the category of spectral spaces and the category of Priestley spaces $\mathcal{PS}$, it follows that the categories **Spec** and $\mathcal{PS}$ are equivalent.

A topological space X is said to be an MV-*space* iff there exists an MV-algebra A such that $\mathcal{F}(A)$ (=the set of prime filters of the MV-algebra A equipped with spectral topology) and X are homeomorphic. It is well known that $\mathcal{F}(A)$ with the specialization order (which coincides with the inclusion between prime filters) forms a root system. Actually any MV-space is a Priestly space which is a root system. An MV-*space* is a Priestley space X such that $R(x)$ is a chain for any $x \in X$ and a morphism between MV-spaces is a strongly isotone map (or an MV-*morphism*), i.e. a continuous map $\varphi : X \to Y$ such that $\varphi(R(x)) = R(\varphi(x))$ for all $x \in X$(for details see [12, 18]). Hence, any MV-space forms a root system. We denote the category of MV-spaces plus MV-morphisms by $\mathcal{MVS}$.

We are interested in subcategory $\mathcal{MVSC}$ of the category $\mathcal{MVS}$, the objects of which are such kind of MV-spaces X for which there exist $MV(C)$-algebras A such that $\mathcal{M}(A) \cong X$, where $\mathcal{M}(A)$ $(= Spec(A))$ is the set of all prime MV-filters.

Notice that the spectral spaces of ℓ-groups (also with strong unit was investigated in [19]), are root systems (or in other terminology, completely normal spectral spaces). Not every completely normal spectral space is a spectral space of some ℓ-group. Notice, also, that there exists an ℓ-group G (with strong unit) such that the distributive lattice, corresponding to the spectral space $Spec(G)$, is not dual Heyting (or op-Heyting) algebra [19]. Taking into account that the category of MV-algebras is equivalent to the category of ℓ-groups with strong unit we conclude that not every

MV-space is a Gödel space. So, more precisely, we are interested in the subcategory of the category $\mathcal{MVSC}$ the objects of which are those $MV(C)$-algebras whose spectral spaces are Gödel spaces.

8.4 Belluce's Functor

On each MV-algebra A, a binary relation $\equiv$ is defined by the following stipulation: $x \equiv y$ iff $supp(x) = supp(y)$, where $supp(x)$ is defined as the set of all prime ideals of A not containing the element x. As proved in [1], $\equiv$ is a congruence with respect to $\oplus$ and $\wedge$. The resulting set $\beta(A)(= A/\equiv)$ of equivalence classes is a bounded distributive lattice, called the Belluce lattice of A. For each $x \in A$ let us denote by $\beta(x)$ the equivalence class of x. Let $f : A \to B$ be an MV-homomorphism. Then $\beta(f)$ is a lattice homomorphism from $\beta(A)$ to $\beta(B)$ defined as follows: $\beta(f)(\beta(x)) = \beta(f(x))$. We stress that β defines a covariant functor from the category of MV-algebras to the category of bounded distributive lattices (see [1]). In [1] (Theorem 20) it is proved that $\mathcal{M}(A)$ and $\mathcal{P}(\beta(A))$ are homeomorphic.

Dually we can define binary relation $\equiv^*$ by the following stipulation: $x \equiv^* y$ iff $supp^*(x) = supp^*(y)$, where $supp^*(x)$ is defined as the set of all prime filters of A containing the element x. Then, $\equiv^*$ is a congruence with respect to $\otimes$ and $\vee$. The resulting set $\beta^*(A)(= A/\equiv^*)$ of equivalence classes is a bounded distributive lattice (which we also call the Belluce lattice of A) $(\beta^*(A), \vee, \wedge, 0, 1)$, where $\beta^*(x) \wedge \beta^*(y) = \beta^*(x \otimes y)$, $\beta^*(x) \vee \beta^*(y) = \beta^*(x \oplus y) = \beta^*(x \vee y)$, $\beta^*(1) = 1$, $\beta^*(0) = 0$, $\beta^*(x)$ is the equivalence class containing the element x. Notice, that if some assertion is true for the functor β, then the same is true for the functor β^*.

Let $f : A \to B$ be an MV-homomorphism. Then $\beta^*(f)$ is a lattice homomorphism from $\beta^*(A)$ to $\beta^*(B)$ defined as follows: $\beta^*(f)(\beta^*(x)) = \beta^*(f(x))$. We stress that β^* defines a covariant functor from the category of MV-algebras to the category of bounded distributive lattices (see [1]). $\mathcal{M}(A)$ and $\mathcal{P}(\beta^*(A))$ are homeomorphic ([1] (Theorem 20)). So, in the sequel we will use notation $\mathcal{P}(A)$ instead of $\mathcal{M}(A)$.

Proposition 8.6 *Let $\{A_i\}_{i\in I}$ be a family of $MV(C)$-algebras such that $\beta^*(A_i)$ is a Gödel algebra for every $i \in I$. Then*

$$\beta^*(\prod_{i\in I} A_i) \cong \prod_{i\in I} \beta^*(A_i).$$

Proof A *product* of a family $(A_i)_{i\in I}$ of objects of a category is an object A together with a family $(\pi_i)_{i\in I}$ of morphisms $\pi_i : A \to A_i$ such that for every object B and every family $(\tau_i)_{i\in I}$ of morphisms $\tau_i : B \to A_i$ there exists a unique morphism $\xi : B \to A$ such that $\pi_i \xi = \tau_i$ for $i \in I$.

It is known that the categorical product in the category of MV-algebras, and in the category of distributive lattices as well, coincides with the direct product.

Let $A = \prod_{i \in I} A_i$ be the product of family $(A_i)_{i \in I}$ of $MV(C)$-algebras such that $\beta^*(A_i)$ is a Gödel algebra for every $i \in I$. Let $(\pi_i)_{i \in I}$ be morphisms (projections) $\pi_i : A \to A_i$. Then $\beta^*(\pi_i) : \beta^*(A) \to \beta^*(A_i)$ will be projections from $\beta^*(A)$ onto $\beta^*(A_i)$.

The set $F_i = \{x \in A : \pi_i(x) = 1\}$ is a filter of MV-algebra A such that $\bigcap_{i \in I} F_i = \{1\}$ and what is more $A/F_i \cong A_i$. $\beta^*(F_i) = \{\beta^*(x) : \beta^*(\pi_i)(\beta^*(x)) = \beta^*(1)\}$ is a lattice filter of $\beta^*(A)$ such that $\bigcap_{i \in I} \beta^*(F_i) = \beta^*(1) = \{1\}$. In other words A $(\beta^*(A))$ is a subdirect product of A_i $(\beta^*(A_i)$. Notice that $\{1\}$ is a filter for every MV-algebra A. Moreover, $\beta^*(1) = [1] = \{1\}$.

Further, according to the construction of the filter F_i we have $A/F_i \cong A_i$. So, $\beta^*(A/F_i) \cong \beta^*(A_i)$ and $\beta^*(A)$ is a subdirect product of $\beta^*(A_i)$.

Let us consider the direct product $\prod_{i \in I} \beta^*(A_i)$ of a family $(\beta^*(A_i))_{i \in I}$. Let $\sigma_i : \prod_{i \in I} \beta^*(A_i) \to \beta^*(A_i)$ be the projection for every $i \in I$. We also have morphisms $\beta^*(\pi_i) : \beta^*(A) \to \beta^*(A_i)$. So, according to the definition of product there exists a unique morphism $\xi : \beta^*(A) \to \prod_{i \in I} \beta^*(A_i)$ such that $\sigma_i \xi = \beta^*(\pi_i)$. We should show that ξ is an isomorphism.

The space $Spec(A)$ of prime filters of A and the space $Spec(\beta^*(A))$ of $\beta^*(A)$ are homeomorphic. At the same time the space $Spec(A_i)$ of prime filters of A_i is homeomorphic to the space $Spec(\beta^*(A_i))$ of prime filters of $\beta^*(A_i)$. Now take a co-product $\coprod_{i \in I} Spec(A_i)$ (in the category of Gödel spaces) which is homeomorphic to the $\coprod_{i \in I} Spec(\beta^*(A_i))$. The $\coprod_{i \in I} Spec(A_i)$ corresponds to the product $\prod_{i \in I} A_i$ (by duality) and $\coprod_{i \in I} Spec(\beta^*(A_i))$ corresponds to the product $\prod_{i \in I} \beta^*(A_i)$ (by duality). It means that $Spec(\prod_{i \in I} A_i)$ is homeomorphic to $Spec(\prod_{i \in I} \beta^*(A_i)$. So, the space of prime filters of A and $\prod_{i \in I} \beta^*(A_i)$ are homeomorphic. It means that ξ is an isomorphism. □

Corollary 8.7 *Let $\{A_i\}_{i \in I}$ be a family of MV-algebras. If $\beta^*(A_i)$ is a Heyting lattice (i.e. for every $x, y \in A_i$ there exists $x \to y$), then $\beta^*(\prod_{i \in I} A_i)$ is also Heyting lattice.*

Proof Since $\beta^*(A_i)$ $(i \in I)$ is a Heyting lattice, we have that $\beta^*(\prod_{i \in I} A_i)$ $(\cong \prod_{i \in I} \beta^*(A_i))$ is also Heyting lattice. □

Proposition 8.8 [1] *Let $\varepsilon : A \to B$ be an injective MV-homomorphism between MV-algebras A and B. Then $\beta^*(\varepsilon) : \beta^*(A) \to \beta^*(B)$ is a distributive lattice injective homomorphism.*

Corollary 8.9 *If A is an MV-subalgebra of MV-algebra B and $\beta^*(B)$ is a Heyting lattice, then $\beta^*(A)$ is also Heyting lattice.*

Proof Let $\varepsilon : A \to B$ is the injective homomorphism corresponding to the subalgebra A of B. Then by [12] (Lemma 13) there exists strongly isotone surjective morphism $\mathcal{P}(f) : \mathcal{P}(B) \to \mathcal{P}(A)$. Therefore, since $\beta^*(B)$ is a Heyting algebra, and so $\mathcal{P}(B)$ is Heyting space, $\mathcal{P}(A)$ is Heyting space and hence A is Heyting algebra. □

8.5 A Weak Duality

Theorem 8.10 *$\beta^*(F_{\mathbf{MV(C)}}(n))$ is a Gödel algebra.*

Proof As we know $F_{\mathbf{MV(C)}}(n) \cong B^{2^n}$, B is a subalgebra of $B_n \times A_1 \times \cdots \times A_{n-1}$ generated by

$$\mathbf{g_1} = (\mathbf{b_1}, \mathbf{d_1^{(1)}}, \ldots, \mathbf{d_1^{(n-1)}}), \ldots, \mathbf{g_n} = (\mathbf{b_n}, \mathbf{d_n^{(1)}}, \ldots, \mathbf{d_n^{(n-1)}}),$$

A_k is a subalgebra of the algebra $\Pi_{i=1}^{\infty} D_i^{(k)}$, where $D_i^{(k)} \cong C_k$ $(1 \leq k < n)$, generated by $\mathbf{d_j^{(k)}} = (u_1^{(j)}, u_2^{(j)}, u_3^{(j)}, \ldots, u_i^{(j)}, \ldots)$, $j = 1, \ldots, n$ where $u_i^{(1)}, \ldots, u_i^{(n)}$ generate $D_i^{(k)}$ and $(u_i^{(1)}, \ldots, u_i^{(n)}) \neq (u_j^{(1)}, \ldots, u_j^{(n)})$ for $i \neq j$.

According to Corollary 8.7 and 8.9 $\beta^*(B_n)$ and $\beta^*(A_k)$ is a Gödel algebra for every $k = 1, .., n-1$. Since β^* commutes with a direct product (Proposition 36), therefore, $\beta^*(B_n \times A_1 \times \cdots \times A_{n-1})$ is also Gödel algebra. Since B is embedded as $MV(C)$-subalgebra into $B_n \times A_1 \times \cdots \times A_{n-1}$, i.e. there exists injective homomorphism $f : B \to B_n \times A_1 \times \cdots \times A_{n-1}$, according to Corollary 8.9, we have that $\beta^*(B)$ is a Gödel algebra. □

Let $\mathbf{MV(C)^G} = \mathbf{LSP}\{C_n : n \in \omega\}$ be the class of algebras generated from $\{C_n : n \in \omega\}$ by the operators of direct products, subalgebras and direct limits. From here we conclude that $F_{\mathbf{MV(C)}}(n) \in \mathbf{MV(C)^G}$. This class is a full subcategory of the category of $MV(C)$-algebras $\mathbf{MV(C)}$. We can consider $\mathbf{MV(C)^G}$ as the category the objects of which are the algebras from $\mathbf{MV(C)^G}$. Taking into account that $\mathbf{GA}$ is locally finite and any algebra can be represented as a direct limit of finitely generated subalgebras, we have that $\mathbf{GA} = \mathbf{LSP}\{\beta^*(C_n) : n \in \omega\}$.

Theorem 8.11 [12] (Theorem 16) *If R_1, R_2 are finite root systems and $f : R_1 \to R_2$ is a strongly isotone map, then there exist $MV(C)$-algebras $A_1, A_2 \in \mathbf{MV(C)^G}$ and an MV-homomorphism $h : A_1 \to A_2$ such that $\mathcal{P}(A_i) \cong R_i$ $i = 1, 2$.*

Theorem 8.12 *There exist contravariant functor $\mathcal{P} : \mathbf{MV(C)^G} \to \mathcal{GS}$ and contravariant functor $\mathcal{H} : \mathcal{GS} \to \mathbf{MV(C)^G}$ such that $\mathcal{H}(\mathcal{P}(A)) \cong A$ for any object $A \in \mathbf{MV(C)^G}$ and $\mathcal{P}(\mathcal{H}(X)) \cong X$ for any object $X \in \mathcal{GS}$, i.e. the functors $\mathcal{P}$ and $\mathcal{H}$ are dense.*

Moreover, the functor $\mathcal{P} : \mathbf{MV(C)^G} \to \mathcal{GS}$ is full, but not faithful and the functor $\mathcal{H} : \mathcal{GS} \to \mathbf{MV(C)^G}$ is faithful, but not full.

Proof First of all recall that a spectral space of an $MV(C)$-algebra A is homeomorphic to the spectral space of the distributive lattice $\beta^*(A)$. Let A be any algebra from $\mathbf{MV(C)^G}$. Then A is isomorphic to the direct limit of a direct system of finitely generated subalgebras $\{A_i, \varphi_{ij}\}$, where A_i is a subdirect product of algebras from the family $\{C_n : n \in \omega\}$ and $\varphi_{ij} : A_i \to A_j$ is an injective homomorphism, $i \leq j$ (more precisely A_i is a subalgebra of A_j). Identify A with its direct limit which is a direct limit of the direct system $\{A_i, \varphi_{ij}\}$. By Corollary 8.9 any $\beta^*(A_i)$ is a Gödel algebra.

By [12] (Theorem 11) we know that β^* preserves direct limits, so, $\beta^*(A)$, which is direct limit of the direct system $\{\beta^*(A_i), \beta^*(\varphi_{ij})\}$ of Gödel algebras, where $\beta^*(\varphi_{ij})$ is a Heyting homomorphism, is also Gödel algebra. We associate the $MV(C)$-space $\mathcal{F}(A) = \mathcal{F}(\beta^*(A))$ to the $MV(C)$-algebra $A \in \mathbf{MV(C)^G}$. Notice that $\mathcal{P}(\beta^*(A))$ is homeomorphic to $\mathcal{F}(\beta^*(A))$. So, we have constructed contravariant functor $\mathcal{P}$ from the category $\mathbf{MV(C)^G}$ to the category of Gödel spaces : $\mathcal{P} : \mathbf{MV(C)^G} \to \mathcal{GS}$.

Let (X, R) be Gödel space. So, a Heyting algebra $\mathcal{H}(X)$, corresponding to the Gödel space (X, R), is a Gödel algebra, say G. It is known that the variety of Gödel algebras is locally finite. Therefore, G is isomorphic to the direct limit of a direct system of finite subalgebras $\{G_i, \psi_{ij}\}$, where $\psi_{ij} : G_i \to G_j$ is an injective homomorphism, $i \leq j$ (more precisely G_i is a subalgebra of G_j), i.e. $G = \varinjlim\{G_i, \psi_{ij}\}$. Identify G with its direct limit. According to the duality between the category of Heyting algebras and the category of Heyting spaces, $X = \mathcal{P}(G)$ is the inverse limit of inverse system $\{\mathcal{P}(G_i), \mathcal{P}(\psi_{ij})\}$, where $\mathcal{P}(G_i)$ is finite root system and $\mathcal{P}(\psi_{ij}) : \mathcal{P}(G_j) \to \mathcal{P}(G_i)$ is a strongly isotone onto map. Then, by [12] (Theorem 15), there exists $MV(C)$-algebras $A_i \in \mathbf{MV(C)^G}$ such that $\mathcal{P}(\beta^*(A_i)) \cong \mathcal{P}(G_i)$ and injective MV-homomorphism $f_{ij} : A_i \to A_j$ such that $\beta^*(A_i) \cong G_i$ for every $i \in I$ and $\mathcal{P}(\beta^*(f_{ij})) = \mathcal{P}(\psi_{ij})$. So, we have a direct system of $MV(C)$-algebras $\{A_i, f_{ij}\}$, where $f_{ij} : A_i \to A_j$ is an injective homomorphism for $i \leq j$. Let A be the direct limit of this direct system. Then $\mathcal{P}(A) \cong \mathcal{P}(G) \cong X$. So, we have constructed a contravariant functor $\mathcal{H}$, such that for a given Gödel space $\mathcal{H}(X) = A$.

From the construction of the functors $\mathcal{P}$ and $\mathcal{H}$ we conclude that $\mathcal{H}(\mathcal{P}(A)) \cong A$ for any object $A \in \mathbf{MV(C)^G}$ and $\mathcal{P}(\mathcal{H}(X)) \cong X$ for any object $X \in \mathcal{GS}$, i.e. the functors $\mathcal{P}$ and $\mathcal{H}$ are dense.

If we have strongly isotone map $f : X_1 \to X_2$ between Gödel spaces X_1 and X_2, then there exist algebras $A_1, A_2 \in \mathbf{MV(C)^G}$ and MV-algebra homomorphism $h : A_2 \to A_1$ such that $\mathcal{P}(A_1) = X_1$, $\mathcal{P}(A_2) = X_2$ (up to isomorphism) and $\mathcal{P}(h) : \mathcal{P}(A_2) \to \mathcal{P}(A_1)$ is strongly isotone. So, $\mathcal{P}$ is full. Now, let us consider two different MV-homomorphisms $f_1, f_2 : C \to C$ such that $f_1(c) = 2c$ and $f_2(c) = 3c$. Nevertheless, $\mathcal{P}(f_1) = \mathcal{P}(f_2) : \mathcal{P}(C) \to \mathcal{P}(C)$. So, $\mathcal{P}$ is not faithfull.

It is obvious that if we have two different morhisms $g_1 : X_1 \to X_2$ and $g_1' : X_1' \to X_2'$, then we have two different MV-homomorphisms $\mathcal{H}(g_1) : \mathcal{H}(X_2) \to \mathcal{H}(X_1)$ and $\mathcal{H}(g_1') : \mathcal{H}(X_2') \to \mathcal{H}(X_1')$. So, $\mathcal{H}$ is faithfull. For the strongly isotone identity map $f : \mathcal{P}(C) \to \mathcal{P}(C)$, we have identity MV-homomorphism from C to C. But for non-trivial injective homomorphism $h : C \to C$, such that $h(c) = 3c$, there is no (not identity) strongly isotone map $g : \mathcal{P}(C) \to \mathcal{P}(C)$ such that $\mathcal{H}(g) = h$. So, $\mathcal{H}$ is not full. □

The category $\mathcal{GS}$ of Gödel spaces is dually equivalent to the category **GA** of Gödel algebras, i.e. there exist two functors $\mathcal{G} : \mathbf{GA} \to \mathcal{GS}$ and $\mathcal{HS} : \mathcal{GS} \to \mathbf{GA}$. So, we have a composition of two contravariant functors $\mathcal{HS} \circ \mathcal{P} : \mathbf{MV(C)^G} \to \mathbf{GA}$ and $\mathcal{H} \circ \mathcal{G} : \mathbf{GA} \to \mathbf{MV(C)^G}$.

From the above, we have the following

Theorem 8.13 *Covariant functors* $\mathcal{HS} \circ \mathcal{P} : \mathbf{MV(C)^G} \to \mathbf{GA}$ *and* $\mathcal{H} \circ \mathcal{G} : \mathbf{GA} \to \mathbf{MV(C)^G}$ *are dense. Moreover,* $\mathcal{HS} \circ \mathcal{P}$ *coincides with Belluce functor* β *defined on the* $\mathbf{MV(C)^G}$.

8.6 Coproduct in $\mathbf{MV(C)^G}$

In this section we will describe finite coproduct $C_1 \sqcup \cdots \sqcup C_1$ (m times) of algebras C_1. Suppose $\mathbf{V}$ is a class of algebras, and $A, B \in \mathbf{V}$. The $\mathbf{V}$-*coproduct* of A and B is an algebra $A \sqcup B \in \mathbf{V}$ with algebra homomorphisms $i_A : A \to A \sqcup B, i_B : B \to A \sqcup \mathbf{B}$, such that $i_A(A) \cup i_B(B) \subset A \sqcup B$ generates $A \sqcup B$, satisfying the following universal property: for every algebra $D \in \mathbf{V}$ with algebra homomorphisms $f : A \to D$ and $g : B \to D$, there exists an algebra homomorphism $h : A \sqcup B \to D$ such that $h \circ i_A = f$ and $h \circ i_B = g$. If we change in the definition of coproduct the requirement that the algebra homomorphisms to be injective, then we have the definition of *free product*. The coproduct $A \sqcup B$ coincides with free product if there is an algebra D such that the algebras A and B can be jointly embedded into D [20]. Since for any $MV(C)$-algebras A and B there is an algebra D such that the algebras can be jointly embedded into D, then the coproduct $A \sqcup B$ in $\mathbf{V}$ coincides with free product. More precisely we have

Theorem 8.14 *In the class* $\mathbf{MV(C)^G}$ *a coproduct coincides with free product.*

Proof Let A, B be any algebras from $\mathbf{MV(C)^G}$. Then $\mathcal{P}(A), \mathcal{P}(B)$, respectively, corresponding to their Gödel spaces. So, since the functor $\mathcal{P}$ is contravariant we have that $\mathcal{P}(A \times B) = \mathcal{P}(A) \uplus \mathcal{P}(B)$ where $A \times B$ is the direct product of A and B, and $\mathcal{P}(A) \uplus \mathcal{P}(B)$ is disjoint union of $\mathcal{P}(A)$ and $\mathcal{P}(B)$. Let a be a maximal element of $\mathcal{P}(A)$ and b a maximal element of $\mathcal{P}(B)$. There exist two different strongly isotone surjective maps $f_A : \mathcal{P}(A) \uplus \mathcal{P}(B) \to \mathcal{P}(A)$ and $f_B : \mathcal{P}(A) \uplus \mathcal{P}(B) \to \mathcal{P}(B)$ such that $f_A(x) = a$ for every $x \in \mathcal{P}(B)$, $f_A(x) = x$ for every $x \in \mathcal{P}(A)$ and $f_B(x) = b$ for every $x \in \mathcal{P}(A)$, $f_B(x) = x$ for every $x \in \mathcal{P}(B)$. So, there exist two injective homomorphisms $\varepsilon_A : A \to A \times B$ and $\varepsilon_B : B \to A \times B$. From here we conclude that the coproduct in $\mathbf{MV(C)^G}$ coincides with free product. □

Let us notice that the coproduct coincides with the free product in the variety of abelian ℓ-groups with strong unit [6].

Now we describe coproduct $C_1 \sqcup C_1$. Recall that finitely generated totally ordered $MV(C)$-algebras are sudirectly irreducible (Lemma 8.5) and observe that the totally ordered $MV(C)$-algebras from $\mathbf{MV(C)^G}$ are C_n, where $n \in Z^+$. In its turn C_1 is generated by one element $c_1 \in C$. Moreover, any element $u(\neq c_1)$ from $Rad(C_1)$ generates a proper subalgebra which is isomorphic to C_1. So, there are infinitely many injective homomorphisms from C_1 into C_1 and for any injective homomorphism $h : C_1 \to C_1$ $h(c_1) = mc_1$ for some $m \in Z^+$. So, if we have two injective homomorphisms $h_1 : C_1 \to C_1$ and $h_2 : C_1 \to C_1$ such that $h_1(c_1) = mc_1$ and $h_2(c_1) = kc_1$, where $m, k \in Z^+$, then $h_1(C_1) \cup h_2(C_1)$ generates C_1 only in the case when m and k are coprime. Now, let us consider injective homomorphisms from C_1 into C_2 that generates C_2. In this case we have only two possibilities $i_1 : C_1 \to C_2$, $i_2 : C_1 \to C_2$ such that $i_1(c_1) = c_1$, $i_2(c_1) = c_2$, and $j_1 : C_1 \to C_2$, $j_2 : C_1 \to C_2$ such that $j_1(c_1) = c_2$, $j_2(c_1) = c_1$.

Now let us consider the algebra $Rad(C_2^2 \times \prod_{i=1}^{\infty} C_1^{(i)}) \cup \neg Rad(C_2^2 \times \prod_{i=1}^{\infty} C_1^{(i)})$, where $C_1^{(i)} \cong C_1^{(j)} \cong C_1$ for any $i, j \in Z^+$. Let $B(2)$ be the subalgebra of $Rad(C_2^2 \times \prod_{i=1}^{\infty} C_1^{(i)}) \cup \neg Rad(C_2^2 \times \prod_{i=1}^{\infty} C_1^{(i)})$ generated by $g_1 = (c_1, c_2, a_1, a_2, \ldots, a_i, ...)$ and $g_2 = (c_2, c_1, b_1, b_2, \ldots, b_i, ...)$, where $(c_1, c_2), (c_2, c_1) \in C_2^2$, $(a_1, a_2, \ldots, a_i, ...), (b_1, b_2, \ldots, b_i, ...) \in \prod_{i=1}^{\infty} C_1^{(i)}$, $a_i = m_i c_1$, $b_i = k_i c_1$ and m_i and k_i are coprime. Notice that $a_i (= m_i c_1)$, $b_i (= k_i c_1)$ generate C_1. Observe, that the algebra $B(2)$, which is a homomorphic image of $F_{\mathbf{MV(C)}}(2)$, is the same (up to isomorphism) which have described in the Sect. 8.3.

It is obvious that the subalgebra of $B(2)$ generated by g_i $(i = 1, 2)$ is isomorphic to C_1. So, we have two injective homomorphisms $i_1 : C_1 \to B(2)$, sending the element c_1 to g_1, and $i_2 : C_1 \to B(2)$, sending the element c_1 to g_2. It is obvious that $i_1(C_1) \cup i_2(C_1)$ generates $B(2)$. Let us suppose that we have algebra $D \in \mathbf{MV(C)^G}$ such that there exist homomorphisms $f : C_1 \to D$ and $g : C_1 \to D$ such that $f(C_1) \cup g(C_1)$ generate D. So, D is generated by $f(c_1), g(c_1) \in D$. It is well known that any algebra is (up to isomorphism) a subdirect product of subdirectly irreducible algebras. Notice that any totally ordered finitely generated $MV(C)$-algebra is subdirectly irreducible. As we know D is a subdirect product of totally ordered $MV(C$-algebras $\pi_i(D)$, which are two-generated, where $\pi_i(D)$ is isomorphic either C_2 or C_1. Therefore, $\pi_i(D)$ is generated by the set $\{\pi_i(f(c_1)), \pi_i(g(c_1))$. Hence, if $\pi_i(D) \cong C_1$, then $\pi_i(f(c_1)) = mc_1$, for some $m \in Z^+$, and $\pi_i(g(c_1)) = kc_1$, for some $k \in Z^+$, and, moreover, since mc_1 and kc_1 generate C_1, we have that m and k are coprime. If $\pi_i(D) \cong C_2$, then either $\pi_i(f(c_1)) = c_1$ and $\pi_i(g(c_1)) = c_2$, or $\pi_i(f(c_1)) = c_2$ and $\pi_i(g(c_1)) = c_1$. So, according to the construction of the algebra $B(2)$, there exists a surjective homomorphism $\tau : B \to D$ such that $\tau \circ i_1 = f$ and $\tau \circ i_2 = g$. From here we arrived to the following

Theorem 8.15 *The algebra $B(2)$ is isomorphic to the coproduct $C_1 \sqcup C_1$.*

We can extend this result on the coproduct $C_1 \sqcup \cdots \sqcup C_1$ (m times). Let B be $MV(C)$-algebra, which is a homomorphic image of $F_{\mathbf{MV(C)}}(m)$, is the algebra that have described in the Sect. 8.3. Then we have

Theorem 8.16 *The algebra $B(m)$ is isomorphic to the coproduct $C_1 \sqcup \cdots \sqcup C_1$ (m times).*

Now we show that C_1 and C_2 are projective algebra.

Theorem 8.17 *The $MV(C)$-algebra C_1 is projective.*

Proof Let us denote the Gödel space $\mathcal{P}(C_1)$ by $(\{a, b\}, \leq)$ where $b < a$. To distinct two 2-element chains we provide the elements by indices. As we know one-generated free algebra is isomorphic to $C_1 \times C_1$. Its Gödel space $\mathcal{P}(C_1 \times C_1) = \mathcal{P}(C_1) \uplus \mathcal{P}(C_1)$ is disjoint union of two 2-element chains, say $\mathcal{P}(C_1) = (\{a_1, b_1\}, \leq_1)$, with $b_1 <_1 a_1$, and $\mathcal{P}(C_1) = (\{a_2, b_2\}, \leq_2)$ with $b_2 <_2 a_2$. Then there exists the injective strongly isotone map $\varepsilon : (\{a, b\}, \leq) \to (\{a_1, b_1\}, \leq_1) \uplus (\{a_2, b_2\}, \leq_2)$ such that $\varepsilon(a) = a_1$ and $\varepsilon(b) = b_1$; and there exists the surjective strongly isotone map $h : (\{a_1, b_1\}, \leq_1) \uplus (\{a_2, b_2\}, \leq_2) \to (\{a, b\}, \leq)$ such that $h(a_1) = h(a_2) = h(b_2) = a$ and $h(b_1) = b$. So, it easy to check that $h\varepsilon = Id$, i.e. that $\mathcal{P}(C_1)$ is a retract of $\mathcal{P}(C_1) \uplus \mathcal{P}(C_1)$. Therefore, according to the duality, there exist injective homomorphism $\mathcal{H}(h) : C_1 \to C_1 \times C_1$ and surjective homomorphism $\mathcal{H}(\varepsilon) : C_1 \times C_1 \to C_1$ (which is really projection) such that $\mathcal{H}(\varepsilon)\mathcal{H}(h) = Id_{C_1}$. Hence C_1 is projective algebra in $\mathbf{MV(C)^G}$. □

From this theorem, as a corollary we have

Corollary 8.18 *The $MV(C)$-algebra $C_1 \sqcup \cdots \sqcup C_1$ (m times) is projective.*

Let $B(2) = C_1 \sqcup C_1$ and $\mathcal{P}(B(2))(= (X_{B(2)}, R))$ its Gödel space. Since $B(2)$ is a perfect algebra, $(X_{B(2)}, R)$ has a greatest element, which we denote by m. Moreover, since $B(2)$ contains infinitely many copies of C_1, we have that $(X_{B(2)}, R)$ contains infinitely many copies of two-element chain up-sets, and two three-element chain up-sets where one of them corresponds to the algebra C_2 with generators $g_1 = c_1$ and $g_2 = c_2$, and the other corresponds to the algebra C_2 with generators $g_1 = c_2$ and $g_2 = c_1$. $(X_{B(2)}, R)$ is depicted in the Fig. 8.1. Notice that the filter F_m generated by $\neg\mathbf{g}_1 \wedge \neg\mathbf{g}_2$ is a maximal prime filter. Moreover, $supp^*(\neg\mathbf{g}_1 \wedge \neg\mathbf{g}_2) = \{F_m\}$. Therefore, $\{F_m\}$ is a clopen.

Let (X, R) be a poset and $x \in X$. A *chain out* of x is a linearly ordered subset (i.e. for every y, z from the subset either yRz or zRy) of X with the least element x; the *depth of* x denotes the supremum cardinality of chains out of x.

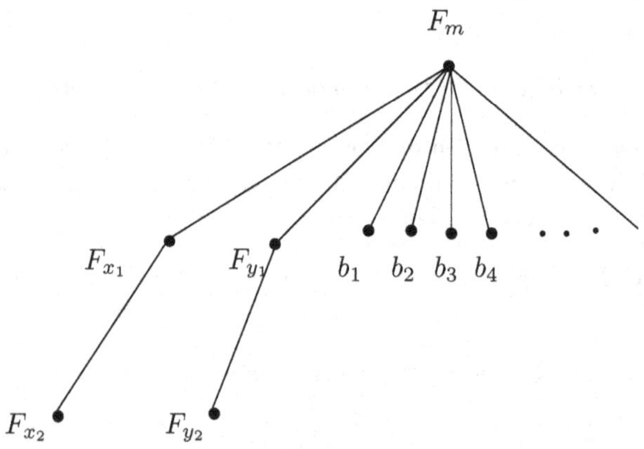

Fig. 8.1 Gödel space $(X_{B(2)}, R)$ of the algebra $B(2)$

Theorem 8.19 *The $MV(C)$-algebra C_2 is projective.*

Proof Let us denote by V_i the set of all elements of $X_{B(2)}$ having the depth not more than i, i.e. the chain $R(x)$ has not more than i elements for any $x \in V_i$. The element $x \in X_{B(2)}$ has the *depth* k if $R(x)$ contains exactly k element. So, $F_m \in X_{B(2)}$ has the depth 1.

Notice that $cl(V_2) = V_2$, where cl is the closure operator of the space $X_{B(2)}$. Indeed, if $cl(V_2) \neq V_2$, then $V_2 \subset cl(V_2)$, which is a dense subset of $cl(V_2)$, contains the elements of depth more than 2. But, according to the duality between Gödel algebras and Gödel spaces, it is impossible, since, in this case, $\mathcal{H}(cl(V_2))$ is isomorphic to a subdirect product of three-element Gödel algebras, that does not contain as a homomorphic image k-element totally ordered Gödel algebra for $k > 2$. From here we conclude that V_2 and $V_2 - \{F_m\}$, as well, are clopen. So, $X_{B(2)} - V_2 = \{F_{x_2}, F_{y_2}\}$ is also clopen. Let $\mathcal{P}(C_2) = (\{F_1, F_2, F_3\}, \subset)$, where F_1 is the prime filter generated by $\neg c_2$, F_2 is the prime filter generated by $\neg c_1$ and F_3 is the prime filter generated by 1. It is obvious that $F_3 \subset F_2 \subset F_1$. Let $\varepsilon : (\{F_1, F_2, F_3\}, \subset) \to (X_{B(2)}, R)$ be the injective strongly isotone map that is defined in the following way: $\varepsilon(F_1) = F_m$, $\varepsilon(F_2) = F_{x_1}$, $\varepsilon(F_3) = F_{y_1}$ and let $h : (X_B, R) \to (\{F_1, F_2, F_3\}, \subset)$ be the continuous surjective strongly isotone map that is defined in the following way: $h(F_m) = F_1$, $h(x) = F_2$ for every $x \in V_2 - V_1$, and $h(x) = F_3$ for every $x \in V_3 - V_2$. It is easy to check that $h\varepsilon = Id$, i.e. $(\{F_1, F_2, F_3\}, \subset)$ is a retract of $(X_{B(2)}, R)$. So, there exist surjective homomorphism $\mathcal{H}(\varepsilon) : B(2) \to C_2$ and injective homomorphism $\mathcal{H}(h) : C_2 \to B$ such that $\mathcal{H}(\varepsilon)\mathcal{H}(h) = Id_{C_2}$. So, C_2 is a retract of $C_1 \sqcup C_1$, i.e. C_2 is projective. □

In the same manner, we can prove the following

Theorem 8.20 *The $MV(C)$-algebra $B_n = Rad(C_n^{n!}) \cup \neg Rad(C_n^{n!})$ is projective for any $n \in Z^+$.*

Proof As in the proof of Theorem 8.19 we denote by V_i the set of all elements of $X_B(= \mathcal{P}(B))$ having the depth not more than i, i.e. the chain $R(x)$ has not more than i elements for any $x \in V_i$. The element $x \in X_{B_n}$ has the *depth* k if $R(x)$ contains exactly k element. So, the greatest element $F_m \in X_B$, which is maximal prime filter generated by $\neg \mathbf{g}_1 \wedge \cdots \wedge \neg \mathbf{g}_n$, has the depth 1. Moreover, $supp(\neg \mathbf{g}_1 \wedge \cdots \wedge \neg \mathbf{g}_n) = \{F_m\}$ and, hence, $\{F_m\}$ is clopen.

$cl(V_2) = V_2$, where cl is the closure operator of the space X_B. Indeed, if $cl(V_2) \neq V_2$, then $V_2 \subset cl(V_2)$, which is a dense subset of $cl(V_2)$, contains the elements of depth more than 2. But, according to the duality between Gödel algebras and Gödel spaces, it is impossible, since, in this case, $\mathcal{H}(cl(V_2))$ is isomorphic to a subdirect product of three-element Gödel algebras, that does not contain as a homomorphic image a k-element totally ordered Gödel algebra for $k > 2$. From here we conclude that V_2 and $V_2 - V_1$, as well, are clopen. In the same manner we can prove that V_i and $V_{i+1} - V_i$ are clopen for any $i \in \{1, \ldots, n-1\}$.

X_B contains as up-set the Gödel space $\mathcal{P}(B_n)$. So there exists continuous injective strongly isotone map $\varepsilon : \mathcal{P}(B_n) \to X_B$. Identifying the corresponding elements, we have that $F_{11}(= F_m)$ is the greatest element of $\mathcal{P}(B_n)$. Let us denote by $F_{k1}, \ldots, F_{kn!}$ the elements of $\mathcal{P}(B_n)$ having the depth k, $k = 2, \ldots, n$ with $F_{ki} \leq F_{(k-1)i}$. So, $F_{k1}, \ldots, F_{kn!} \in V_k - V_{k-1}$. Since X_B is a Stone space, i.e. zero-dimensional, compact and Hausdorff, there exists disjoint clopen subsets $U_{21}, \ldots, U_{2n!} \subset V_2 - V_1$ such that $U_{21} \cup \cdots \cup U_{2n!} = V_2 - V_1$ and $F_{2j} \in U_{2j}$ for $j = 1, \ldots, n!$. Let $U_{kj} = (V_k - V_{k-1}) \cap R^{-1}(U_{2j})$ for $j = 1, \ldots, n!$. Then the map $f : X_B \to \mathcal{P}(B_n)$ defined in the following way: $f(x) = F_{ki}$ for every $x \in U_{ki}$, will be continuous strongly isotone. It is easy to check that $f\varepsilon = Id$. Therefore $\mathcal{H}(\varepsilon)\mathcal{H}(f) = Id_{B_n}$. So, B_n is projective. □

It is easy to show that any homomorphic image of the projective algebra $Rad(C_n^{n!}) \cup \neg Rad(C_n^{n!})$ is a retract of $Rad(C_n^{n!}) \cup \neg Rad(C_n^{n!})$, i.e. we have

Corollary 8.21 *Any homomorphic image of the $MV(C)$-algebra* $B_n = Rad(C_n^{n!}) \cup \neg Rad(C_n^{n!})$ *is projective for any $n \in Z^+$. In other words, the algebra $A = C_{n(1)} \times \cdots \times C_{n(k)}$, where $n(1), \ldots, n(k) \leq n$ are positive integers, is projective.*

Theorem 8.22 *Let $A \in \mathbf{MV(C)}^{\mathbf{G}}$. If $\mathcal{P}(A)$ is finite, then A is projective in $\mathbf{MV(C)}^{\mathbf{G}}$.*

Proof Let $A \in \mathbf{MV(C)}^{\mathbf{G}}$ and $\mathcal{P}(A)$ is finite. Then $\mathcal{P}(A)$ is an up-set of $\mathcal{P}(F_{\mathbf{MV(C)}^{\mathbf{G}}}(n))$ for some $n \in Z^+$. Let X be arbitrary root from$\mathcal{P}(F_{\mathbf{MV(C)}^{\mathbf{G}}}(n))$,

i.e. $X \cong R^{-1}(m)$ for some maximal element $m \in \mathcal{P}(F_{\mathbf{MV(C)}^{\mathbf{G}}}(n))$. Notice, that any root of $\mathcal{P}(F_{\mathbf{MV(C)}^{\mathbf{G}}}(n))$ is clopen of $\mathcal{P}(F_{\mathbf{MV(C)}^{\mathbf{G}}}(n))$. So, $X \cap \mathcal{P}(A)$ is closed in general, but the singleton containing the top element of $\mathcal{P}(A)$ is clopen. Let $a_1, \ldots, a_k$ are all elements of $X \cap \mathcal{P}(A)$ having depth 2. Then there exist disjoint clopen sets $U_1, \ldots, U_k \subset V_2 - V_1 \subset X$ such that $U_1 \cup \cdots \cup U_k = V_2 - V_1$ and $a_i \in U_i$ for $i = 1, \ldots, k$. Now let a_i cover the elements $b_1, \ldots, b_m$. Then $R^{-1}(U_i) \cap V_3$ is a clopen and $b_1, \ldots, b_m \in R^{-1}(U_i) \cap V_3$, and, as in the previous case for the elements having the depth 2, there exist disjoint clopen sets $W_1, \ldots, W_m$ such that $\bigcup_{i=1}^m W_i - R^{-1}(U_i) \cap V_3$ and $b_i \in W_i$. The same procedure we make for elements $x \in \mathcal{P}(A)$ having depth more than 3 and so on. Let y be a bottom element of $\mathcal{P}(A)$ having the depth k and $Y \subset V_k$ the clopen containing the element y. Let $R^{-1}(Y)$ be the class containing the element y. So, we have finite partition of $\mathcal{P}(F_{\mathbf{MV(C)}^{\mathbf{G}}}(n))$ on clopen classes such that for every $x \in \mathcal{P}(A)$ we have clopen, say U_x, and if $x \neq y$, then $U_x \neq U_y$. Moreover, if V is an upper set of $\mathcal{P}(A)$, then $\bigcup\{U_x : x \in V\}$ is a clopen upper set of $\mathcal{P}(F_{\mathbf{MV(C)}^{\mathbf{G}}}(n))$. Let $f : \mathcal{P}(F_{\mathbf{MV(C)}^{\mathbf{G}}}(n)) \to \mathcal{P}(A)$ be the map such that $f(y) = x$ if $y \in V_x$, where V_x is an element of the partition containing the element x. It is obvious that f is strongly isotone. So we have injective continuous strongly isotone map $\varepsilon : \mathcal{P}(A) \to \mathcal{P}(F_{\mathbf{MV(C)}^{\mathbf{G}}}(n))$ and surjective continuous strongly isotone map $f : \mathcal{P}(F_{\mathbf{MV(C)}^{\mathbf{G}}}(n)) \to \mathcal{P}(A)$ such that $f\varepsilon = Id_{\mathcal{P}(A)}$. Therefore, $\mathcal{H}(\varepsilon)\mathcal{H}(f) = Id_A$, i.e. A is projective. □

Recall that an MV-algebra A is *finitely presented* iff $A \cong F_{\mathbf{MV}}(m)/[u)$ for some principal filter generated by $u \in F_{\mathbf{MV}}(m)$ [21, 22].

Theorem 8.23 *Any finitely presented algebra $A \in \mathbf{MV(C)^G}$ is projective.*

Proof Let $A \in \mathbf{MV(C)^G}$ be finitely presented. Then it is n-generated for some $n \in Z^+$. Hence, it is a homomorphic image of $F_{\mathbf{MV(C)^G}}(n)$. Moreover, there exists principal filter $[u)$ for some $u \in F_{\mathbf{MV(C)^G}}(n)$ such that $A \cong F_{\mathbf{MV(C)^G}}(n)/[u)$. It means that there exists continuous strongly isotone map $\varepsilon : \mathcal{P}(A) \to \mathcal{P}(F_{\mathbf{MV(C)^G}}(n))$ such that $\varepsilon(\mathcal{P}(A))$ is a clopen of $\mathcal{P}(F_{\mathbf{MV(C)^G}}(n))$, which corresponds to the element $\beta^*(u) \in \beta^*(\mathcal{P}(F_{\mathbf{MV(C)^G}}(n)))$.

Notice that the root system $\mathcal{P}(F_{\mathbf{MV(C)^G}}(n))$ consists of 2^n roots (that are isomorphic to each other). Partition every root on closed classes in such a way that any class contains only one element from $\varepsilon(\mathcal{P}(A))$. Let X be arbitrary root from $\mathcal{P}(F_{\mathbf{MV(C)^G}}(n))$. Notice, that any root of $\mathcal{P}(F_{\mathbf{MV(C)^G}}(n))$ is clopen of $\mathcal{P}(F_{\mathbf{MV(C)^G}}(n))$. So, $X \cap \mathcal{P}(A)$ is clopen. Since $\varepsilon(\mathcal{P}(A))$ is a clopen of $\mathcal{P}(F_{\mathbf{MV(C)^G}}(n))$, we have that $V_2' = V_2 \cap \varepsilon(\mathcal{P}(A))$ is a clopen of X consisting of the elements of $\varepsilon(\mathcal{P}(A))$ having the depth 2. Let $X - R^{-1}(V_2')$ be the class (that is clopen) which contains the only maximal element, say F_m, of X belonging to $\varepsilon(\mathcal{P}(A))$ and, moreover, $X - R^{-1}(V_2')$ is a clopen upset. Notice, that since $\varepsilon(\mathcal{P}(A))$ is a clopen in $\mathcal{P}(F_{\mathbf{MV(C)^G}}(n))$, we have that $V_i' = V_i \cap \varepsilon(\mathcal{P}(A)) \subset X$ is also clopen in $\mathcal{P}(F_{\mathbf{MV(C)^G}}(n))$. Therefore, the set of minimal elements $Min(\varepsilon(\mathcal{P}(A)))$ of $\varepsilon(\mathcal{P}(A))$ is also clopen in $\mathcal{P}(F_{\mathbf{MV(C)^G}}(n))$. Now let us suppose that $t \in \varepsilon(\mathcal{P}(A))$ is not minimal element of $\varepsilon(\mathcal{P}(A))$ and have depth $k > 2$. Let $R^{-1}(t) - V_{k+1}'$ be the class which contains the only element t from $\varepsilon(\mathcal{P}(A))$. It is clear that $R^{-1}(t) - V_{k+1}'$ is closed (clopen) if $\{t\}$ is closed (clopen). So, $R^{-1}(Min(\varepsilon(\mathcal{P}(A))))$ is clopen. The following classes of our needed partition are $R^{-1}(t)$ for every minimal element $t \in Min(\varepsilon(\mathcal{P}(A)))$. Notice that if $\{t\}$ is clopen, then $R^{-1}(t)$ is clopen and if $\{t\}$ is closed, then $R^{-1}(t)$ is closed. Moreover, the class $R^{-1}(t)$ contains the only element t belonging to $\varepsilon(\mathcal{P}(A))$. So, we have a correct partition [12] of X and, hence, $\mathcal{P}(F_{\mathbf{MV(C)^G}}(n))$. It means that for any closed upset U of $\mathcal{P}(F_{\mathbf{MV(C)^G}}(n))$ the saturation $E(U) = \bigcup_{x \in U} E(x)$ is also closed, and if U is clopen of $\varepsilon(\mathcal{P}(A))$, then $E(U) = \bigcup_{x \in U} E(x)$ is clopen of $\mathcal{P}(F_{\mathbf{MV(C)^G}}(n))$, where E is an equivalence relation corresponding to the constructed partition and $E(x) = \{y : xEy\}$.

From the above, we conclude that there exists surjective continuous strongly isotone map $f : \mathcal{P}(F_{\mathbf{MV(C)^G}}(n)) \to \mathcal{P}(A)$ such that $f(x) = \varepsilon^{-1}(y)$ where $\{y\} = E(x) \cap \varepsilon(\mathcal{P}(A))$, i.e. y is the only element of $E(x)$ belonging to $\varepsilon(\mathcal{P}(A))$. It is easy to check that $f\varepsilon = Id_{\mathcal{P}(A)}$, i.e. $\mathcal{P}(A)$ is a retract of $\mathcal{P}(F_{\mathbf{MV(C)^G}}(n))$. From here we deduce that $\mathcal{H}(\varepsilon)\mathcal{H}(f) = Id_A$ and, hence, A is a projective algebra. □

We selected the class $\mathbf{MV(C)^G}$ from the variety $\mathbf{MV(C)}$ generated by perfect MV-algebras. This class is formed by the operators of taking direct products, subalgebras and direct limits on the set $\{C_i : 0 \leq i < \omega\}$, where C_i is i-generated perfect totally ordered MV-algebra. The class $\mathbf{MV(C)^G}$ forms a full subcategory of $\mathbf{MV(C)}$. It is well known that Gödel algebras, that is Heyting algebras with linearity condition, are dually equivalent to the category $\mathcal{GS}$ of Heyting spaces of the Gödel algebras.

In this chapter, we constructed two functors $\mathcal{P} : \mathbf{MV(C)^G} \to \mathcal{GS}$ and $\mathcal{H} : \mathcal{GS} \to \mathbf{MV(C)}$ such that $\mathcal{P}$ is full and $\mathcal{H}$ is faithful, and both functors are dense. That is we proved that the categories $\mathbf{MV(C)^G}$ and $\mathcal{GS}$ are weakly dual.

Also the description of finite coproduct of Chang's algebras is given, and using the above weak duality, a characterization of projective algebras is given too.

References

1. Belluce, L.P.: Semisimple algebras of in infinite-valued logic and bold fuzzy set theory. Canad. J. Math. **38**, 1356–1379 (1986)
2. Holland, W.C., Scrimger, E.: Free product of lattice-ordered groups. Algebra Univ. **2**, 247–254 (1972)
3. Martinez, J.: Free products in varieties of lattice-ordered groups. Czechoslov. Math. J. **22**(97), 535–553 (1972)
4. Martinez, J.: Free products of abelian ℓ-groups. Czechoslov. Math. J. **23**(98), 349–361 (1973)
5. Powell, W.B., Tsinakis, C.: Free products in varieties lattice-ordered groups. In: Glass, A.M.W., Holland, W.C. (eds.) Lattice-Ordered Groups, pp. 278.307. D. Reidel, Dordrecht (1989)
6. Mundici, D.: The Haar theorem for lattice-ordered abelian groups with order-unit. Discret. Contin. Dyn. Syst. **21**, 537–549 (2008)
7. Dvureaenskij, A., Holland, W.C.: Free products of unital ℓ-groups and free products of generalized MV-algebras. Algebra Univers. **62**(1), 19–25 (2009)
8. Di Nola, A., Lettieri, A.: Perfect MV-algebras are categorically equivalent to abelian ℓ-groups. Stud. Logica **88**, 467–490 (1994)
9. Di Nola, A. Grigolia, R.: Finiteness and duality in MV-algebras theorry, Advances in Soft Computing, Lectures on Soft Computing and Fuzzy Logic, pp. 71–88. Physica-Verlag, A Springer-Verlag Company (2001)
10. Pogorzelski, W.A.: Structural completeness of the propositional calculus. Bull. Acad. Polon. Sci., Ser. Math. Astr. Phys. **19**, 349–351 (1971)
11. Davey, B.A., Priestley, H.A.: Introduction to Lattices and Order, 2nd edn. Cambridge University Press, Cambridge (2002)
12. Di Nola, A. Grigolia, R.: Profinite MV-spaces. Discret. Math. **283**(1–3), 61–69 (2004)
13. Horn, A.: Logic with truth values in a linearly ordered heyting algebra. J. Symbo. logic **34**, 395–408 (1969)
14. Panti, G.: Varieties of MV-algebras. J. Appl. Non-Class. Logics **9**(1), 141–157 (1999)
15. Chang, C.C.: Algebraic analysis of many-valued logics. Trans. Am. Math. Soc. **88**, 467–490 (1958)
16. Birkhoff, G.: Lattice Theory. Providence, Rhode Island (1967)
17. Burris, S., Sankappanavar, H.P.: A Course in Universal Algebras. The Millenium Edition (2000)
18. Di Nola, A., Grigolia, R.: MV-algebras in duality with labeled root systems. Discret. Math. **243**, 79–90 (2002)
19. Cignoli, R., Glushankof, D., Lucas, F.: Prime spectra of lattice-ordered abelian groups. J. Pure Appl. Algebra **136**, 217–229 (1999)
20. Malcev, A.I.: Algebraic Systems. Springer (1973). ISBN 0-387-05792-7
21. Cignoli, R.L.O., D'Ottaviano, I.M.L., Mundici, D.: Algebraic Foundations of Many-Valued Reasoning. Trends in Logic|Studia Logica Library, vol. 7. Kluwer Academic Publishers, Dordrecht (2000)
22. Di Nola, A, Grigolia, R.: Projective MV-algebras and their automorphism groups. J Multi-Valued Logic Soft Comput. **9**, 291–317 (2015)

Chapter 9
The Logic of Perfect Algebras

As we know the MV-algebra C is the simplest MV-algebra with infinitesimals. That is, any non semisimple MV-algebra contains a copy of C as subalgebra. C is generated by an atom c, which we can interpret as a quasi false truth value. The negation of c is a quasi true value. Now, quasi truth or quasi falsehood are vague concepts. Hence, it is quite intriguing to explore such a logic of *quasi true*. About quasi truth in an MV algebra, it is reasonable to accept the following propositions:

- there are quasi true values which are not 1;
- 0 is not quasi true;
- if x is quasi true, then x^2 is quasi true (where x^2 denotes the MV-algebraic product of x with itself).

In C, to satisfy these axioms, it is enough to say that the quasi true values are the co-infinitesimals.

By way of contrast, note that there is no notion of quasi truth in [0, 1] satisfying the previous axioms (there are if we replace the MV product with other suitable t-norms, e.g. the product t-norm or the minimum t-norm).

Recall that algebras from the variety generated by C will be called by $MV(C)$-*algebra*. Also we recall that for an $MV(C)$-algebra A, its *Boolean skeleton*, $B(A)$, that is the greatest Boolean subalgebra of A, is a retract of A, via the radical ideal of A, see [1]. Thus, roughly speaking, every $MV(C)$-algebra can be seen as a Boolean algebra, up to infinitesimals.

Let L_P be the logic corresponding to the variety generated by perfect algebras which coincides with the set of all Łukasiewicz formulas that are valid in all perfect MV-chains, or equivalently that are valid in the MV-algebra C. Actually, L_P is the logic obtained by adding to the axioms of Łukasiewicz sentential calculus the following axiom: $(x \oplus x) \odot (x \oplus x) \leftrightarrow (x \odot x) \oplus (x \odot x)$, see [2]. Notice that the above axiom is used in [3] to define an interesting class of Glivenko MTL-algebras and that the Lindenbaum algebra of L_P is an $MV(C)$-algebra.

A. Di Nola et al., *Fuzzy Logic of Quasi-Truth: An Algebraic Treatment*,
Studies in Fuzziness and Soft Computing 338, DOI 10.1007/978-3-319-30406-9_9

The importance of the class $MV(C)$-algebras and of the logic L_P can be percieved looking further at the role that infinitesimals play in MV-algebras and in Łukasiewicz logic. Indeed the pure first order Łukasiewicz predicate logic is not complete with respect the canonical set of truth values [0, 1] [4]. However a completeness theorem is obtained if the truth values are allowed to vary through all linearly ordered MV-algebra [5]. From the incompleteness theorem arises the problem of the algebraic significance of the true but unprovable formulas. In [6] it is remarked that the Lindenbaum algebra of first order Łukasiewicz logic is not semisimple and that the valid but unprovable formulas are precisely the formulas whose negations determine the radical of the Lindenbaum algebra, that is the co-infinitesimals of such algebra. Hence, the valid but unprovable formulas generate the prefect skeleton of the Lindenbaum algebra. So, perfect MV-algebras, the variety generated by them and their logic are intimately related with a crucial phenomenon of first order Łukasiewicz logic.

As it is well known, MV-algebras form a category which is equivalent to the category of abelian lattice ordered groups (ℓ-groups, for short) with strong unit [7]. We denote by Γ the functor implementing this equivalence. In particular each perfect MV-algebra is associated with an abelian ℓ-group with a strong unit. Moreover, the category of perfect MV-algebras is equivalent to the category of abelian ℓ-groups, see [1]. Among perfect MV-algebras the algebra C plays a very important role. Indeed it is the generator of the variety **MV(C)**, the logic L_P is complete with respect to C, and C corresponds to the Behncke-Leptin C^*-algebra $A_{1,0}$ with a two-point dual, via the composition of the functor Γ with K_0, see [8].

From above it is clear that the class of $MV(C)$-algebras, far from being a quite narrow and exotic class it deserves to be explored because of its several and fruitful links with other areas of Logic and Algebra. Now we are going to focus on the logic L_P and especially on its derivability properties.

Derivable and admissible rules were introduced by Lorenzen [9]. A rule

$$\varphi_1, \ldots, \varphi_n / \psi$$

is derivable if it belongs to the consequence relation of the logic (defined semantically, or by a proof system using a set of axioms and rules); and it is admissible if the set of theorems of the logic is closed under the rule. These two notions coincide for the standard consequence relation of classical logic, but nonclassical logics often admit rules which are not derivable. A logic whose admissible rules are all derivable is called *structurally complete*.

Ghilardi [10, 11] discovered the connection of admissibility to projective formulas and unification, which provided another criteria for admissibility in certain modal and intermediate logics (=extensions of intuitionistic logic), and new decision procedures for admissibility in some modal and intermediate logics.

Moreover, following Ghilardi [12] defining unification problem in terms of finitely presented algebras, and having our result that finitely generated finitely presented algebras are precisely finitely generated projective algebras, we deduce that the equational class of all $MV(C)$-algebras has unitary unification type, i.e. L_P has unitary unification type.

Now we give assertions concerning to the completeness of the logic L_P which is the logic corresponding to the variety generated by perfect algebras which coincides with the set of all Łukasiewicz formulas that are valid in all perfect MV-chains.

Theorem 9.1 *A well formed formula α of L_P is valid in algebra C if and only if it is a theorem of L_P.*

Proof Notice that the algebra C generate the variety $\mathbf{MV(C)}$ generated by all perfect MV-algebras. So, we have $C \models p = q$ if and only if $\mathbf{MV(C)} \models p = q$ for any identity $p = q$. Any identity $p = q$ for MV-algebras can be represented as the equivalent one $p \leftrightarrow q = 1$. Therefore, considering any formula α as an algebraic polynomial we can assert that $C \models \alpha = 1$ if and only if $\mathbf{MV(C)} \models \alpha = 1$. From here we conclude that α is valid in algebra C if and only if it is a theorem of L_P. □

Let $Lind_P$ denote the Lindenbaum algebra of the logic L_P. Then we have the following completeness theorem (see [2]).

Theorem 9.2 *A well formed formula of L_P is valid on all prefect MV-chains if and only if it is provable in L_P.*

Proof It is easy to see that if α is a theorem in L_P then α is valid on all perfect MV-algebras. Indeed axioms of L_P are valid in all perfect MV-algebras and modus ponens keeps this validity.

Conversely, $Lind_P$ satisfies $([\alpha]\odot[\alpha])\oplus([\alpha]\odot[\alpha]) = ([\alpha]\oplus[\alpha])\odot([\alpha]\oplus[\alpha])$, that is, $Lind_P \in \mathcal{V}(C)$. Now, let α be a wff of L_P and suppose that α is valid on all perfect MV-chains. Suppose that α is not provable in L_P; then $[\alpha] \neq 1$, and so $[\neg\alpha] \neq 0$. Since $Lind_P$ is semi-perfect there is a prime ideal J such that $\frac{[\neg\alpha]}{J}$. Moreover J is a perfect ideal. So in $\frac{Lind_P}{J}$ we have that $\frac{[\neg\alpha]}{J} \neq 0$ that is $\frac{[\alpha]}{J} \neq 1$. From this we may infer that α is not valid on the perfect MV-chain $\frac{Lind_P}{J}$ via the assignment $v \to \frac{[v]}{J}$ for each propositional variable v. □

Corollary 9.3 *The logic L_P is complete with respect to all ultrapowers of $\Gamma(Z \times_{lex} \mathbb{R}, (1, 0))$, i.e., to all perfect MV-chains of type ${}^*\Gamma(Z \times_{lex} \mathbb{R}, (1, 0))$.*

9.1 Finitely Generated Projective $MV(C)$-Algebras

Definition 9.4 A subalgebra A of $F_\mathbf{V}(m)$ is said to be projective subalgebra if there exists an endomorphism $h : F_\mathbf{V}(m) \to F_\mathbf{V}(m)$ such that $h(F_\mathbf{V}(m)) = A$ and $h(x) = x$ for every $x \in A$.

Proposition 9.5 [13, 14] *Let $\mathbf{V}$ be a variety and $F_\mathbf{V}(m)$ an m-generated free algebra of the variety $\mathbf{V}$, and let $g_1, \ldots, g_m$ be its free generators. Then an m-generated subalgebra A of $F_\mathbf{V}(m)$ with the generators $a_1, \ldots, a_m \in A$ is projective iff there* *exist polynomials* $p_1(x_1, \ldots, x_m), \ldots, p_m(x_1, \ldots, x_m)$ *such that*

$$p_i(g_1, \ldots, g_m) = a_i$$

and

$$p_i(p_1(x_1, \ldots, x_m), \ldots, p_m(x_1, \ldots, x_m)) = p_i(x_1, \ldots, x_m),\ i = 1, \ldots, m,$$

hold in $\mathbf{V}$.

From the Proposition we obtain that in $F_{\mathbf{V}}(m)$ holds

$$p_i(p_1(g_1, \ldots, g_m), \ldots, p_m(g_1, \ldots, g_m)) = p_i(g_1, \ldots, g_m) = a_i,$$

$i = 1, \ldots, m$, i.e. $p_i(a_1, \ldots, a_m) = a_i$ in A. This suggests to consider the free object $F_{\mathbf{V}}(m, \Omega)$ over the variety $\mathbf{V}$ with respect to the set of identities $\Omega = \{p_1(x_1, \ldots, x_m) = x_1, \ldots, p_1(x_1, \ldots, x_m) = x_m\}$.

Proposition 9.6 [13, 15] (Lemmas 2, 3) *An* MV*-algebra* A *is finitely presented iff* $A \cong F_{\mathbf{MV}}(m)/[u)$, *where* $[u)$ *is a principal filter generated by some element* $u \in F_{\mathbf{MV}}(m)$.

Theorem 9.7 *Let* A *be an* m*-generated* $MV(C)$*-algebra. Then the following are equivalent:*

1. *A is projective.*
2. *A is finitely presented.*

Proof 1 $\Rightarrow$ 2. Since A is m-generated projective $MV(C)$-algebra, A is a retract of $F_{\mathbf{MV(C)}}(m)$, i.e. there exist homomorphisms $h : F_{\mathbf{MV(C)}}(m) \to A$ and $\varepsilon : A \to F_{\mathbf{MV(C)}}(m)$ such that $h\varepsilon = Id_A, h(g_i) = a_i$ $(i = 1, \ldots, m)$, and moreover, according to Proposition 8.8, there exist m polynomials $p_1(x_1, \ldots, x_m), \ldots, p_m(x_1, \ldots, x_m)$ such that

$$p_i(g_1, \ldots, g_m) = \varepsilon(a_i) = \varepsilon h(g_i)$$

and

$$p_i(P_1(x_1, \ldots, x_m), \ldots, p_m(x_1, \ldots, x_m)) = p_i(x_1, \ldots, x_m),\ i = 1, \ldots, m,$$

where $g_1, \ldots, g_m$ are free generators of $F_{\mathbf{MV(C)}}(m)$.

Observe that $h(g_1), \ldots, h(g_m)$ are generators of A which we denote by $a_1, \ldots, a_m$ respectively. Let e be the endomorphism $\varepsilon h : F_{\mathbf{MV(C)}}(m) \to F_{\mathbf{MV(C)}}(m)$. This endomorphism has the properties: $ee = e$ and $e(x) = x$ for every $x \in \varepsilon(A)$.

Let us consider the set of identities $\Omega = \{p_i(x_1, \ldots, x_m) \leftrightarrow x_i = 1 : i = 1, \ldots, m\}$ and let $u = \bigwedge_{i=1}^{n}(p_i(g_1, \ldots, g_m) \leftrightarrow g_i) \in F_{\mathbf{MV(C)}}(m)$, where $x \leftrightarrow$

y is abbreviation of $(x \to y) \wedge (y \to x)$. Then, according to Proposition 9.5, $F_{\mathbf{MV(C)}}(m)/[u) \cong F_{\mathbf{MV(C)}}(m, \Omega)$. Observe that the identities from Ω are true in A on the elements $\varepsilon(a_i) = e(g_i)$, $i = 1, \ldots, m$. Indeed, since e is an endomorphism

$$e(u) = \bigwedge_{i=1}^{m} e(g_i) \leftrightarrow p_i(e(g_1), \ldots, e(g_m)).$$

But

$$\begin{aligned} p_i(e(g_1), \ldots, e(g_m)) &= p_i(p_1(g_1, \ldots, g_m), \ldots, p_n(g_1, \ldots, g_m)) \\ &= p_i(g_1, \ldots, g_m) \\ &= \varepsilon h(g_i) \\ &= e(g_i), i = 1, \ldots, m. \end{aligned}$$

Hence $e(u) = 1$ and $u \in e^{-1}(1)$, i.e. $[u) \subseteq e^{-1}(1)$. Therefore there exists a homomorphism $f : F_{\mathbf{MV(C)}}(m)/[u) \to \varepsilon(A)$ such that the diagram

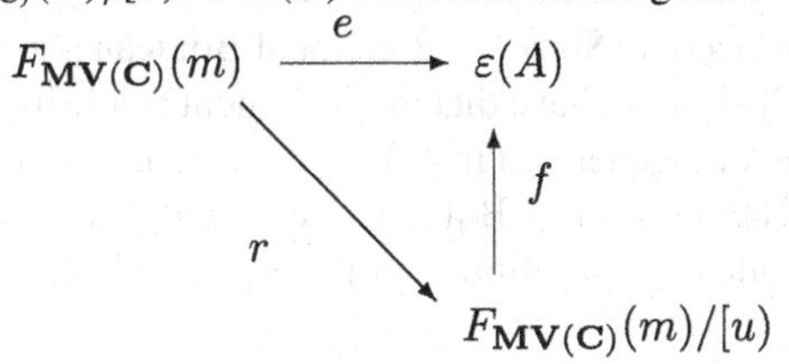

commutes, i.e. $fr = e$, where r is a natural homomorphism sending x to $x/[u)$. Now consider the restrictions e' and r' on $\varepsilon(A) \subseteq F_{\mathbf{MV(C)}}(m)$ of e and r respectively. Then $fr' = e'$. But $e' = Id_{\varepsilon(A)}$. Therefore $fr' = Id_{\varepsilon(A)}$. From here we conclude that r' is an injection. Moreover r' is a surjection, since $r(\varepsilon(a_i)) = r(g_i)$. Indeed $e(g_i) = p_i(g_1, \ldots, g_n)$ and $g_i \leftrightarrow p_i(g_1, \ldots, g_n) = g_i \leftrightarrow e(g_i)$, where $e(g_i) = \varepsilon h(g_i)$. So $g_i \leftrightarrow p_i(g_1, \ldots, g_m) \geq \bigwedge_{i=1}^{m} g_i \leftrightarrow p_i(g_1, \ldots, g_m)$, i.e. $g_i \leftrightarrow p_i(g_1, \ldots, g_m) \in [u)$. Hence r' is an isomorphism between $\varepsilon(A)$ and $F_{\mathbf{MV(C)}}(m)/[u)$. Consequently $A (\cong \varepsilon(A))$ is finitely presented.

$2 \Rightarrow 1$. Let A be an m-generated finitely presented $MV(C)$-algebra. Then there exists a principal filter $[u)$ of m-generated free $MV(C)$-algebra $F_{\mathbf{MV(C)}}(m)$ such that $A \cong F_{\mathbf{MV(C)}}(m)/[u)$ (Proposition 2.3). Since $F_{\mathbf{MV(C)}}(m)$ is a subdirect product of finitely generated chain $MV(C)$-algebras, the element $u \in F_{\mathbf{MV(C)}}(m)$ we can represent as a sequence $(u_i)_{i \in I}$. Let $J = \{i \in I : u_i \neq 1\}$. Let π_J be a natural homomorphism such that $\pi_J((a_i)_{i \in I}) = (a_i)_{i \in J}$. On the other hand the subalgebra of $F_{\mathbf{MV(C)}}(m)$ generated by $[u)$, which is a perfect MV-algebra $[u) \cup \neg[u)$, is isomorphic to $\pi_J(F_{\mathbf{MV(C)}}(m)) \cong F_{\mathbf{MV(C)}}(m)/[u) \cong A$. Notice, that if $(x_i)_{i \in I} \in [u)$, then $x_i = 1$ for $i \in I - J$; and if $(x_i)_{i \in I} \in \neg[u)$, then $x_i = 0$ for $i \in I - J$. So, the set $A' = \{(x_i)_{i \in J} : (x_i)_{i \in I} \in [u) \cup \neg[u)\}$ forms an $MV(C)$-algebra which is isomorphic to $[u) \cup \neg[u)$. Let $\varepsilon : A' \to F_{\mathbf{MV(C)}}(m)$ be the embedding such that $\varepsilon((x_i)_{i \in J}) = (x_i)_{i \in I} \in F_{\mathbf{MV(C)}}(m)$, where $x_i = 1$ if $(x_i)_{i \in J}$ belongs to the maximal filter and $i \in I - J$; and $x_i = 0$ if $(x_i)_{i \in J}$ belongs to the maximal ideal and $i \in I - J$. Thus we conclude that $\pi_J\varepsilon = Id_{A'}$. From here we deduce that the $MV(C)$-algebra A is projective. □

Observe, that for ℓ-groups, Baker [16] and Beynon [17] gave the following characterization: An ℓ-group G is finitely generated projective iff it is finitely presented. For unital ℓ-groups the ($\Rightarrow$)-direction holds [18] (Proposition 2.5). Theorem 8.23 establishes the equivalence for variety of $MV(C)$-algebras.

The algebra C is isomorphic to $\Gamma(Z \times_{lex} Z, (1, 0))$, with generator c (=(0, 1)). In another notation the algebra C is denoted by $S_1^{\omega}(= \Gamma(Z \times_{lex} Z, (1, 0)))$. Recall that **MV(C)** is the variety generated by perfect algebras.

Recall that a 1-generated free $MV(C)$-algebra $F_{\mathbf{MV(C)}}(1)$ is isomorphic to C^2 with free generator $(c, \neg c)$ (Theorem 8.1).

Theorem 9.8 *The two-element Boolean algebra and the $MV(C)$-algebra C are projective.*

Proof It is obvious that the two-element Boolean algebra is projective. Indeed, as we already stressed, the Boolean skeleton $B(C^2)$ is a retract of C^2 [1]. So, the 4-element Boolean algebra is projective. Since the 2-element Boolean algebra is a retract of the 4-element Boolean algebra, we have that the 2-element Boolean algebra is projective. As we know C^2 is the one-generated free $MV(C)$-algebra. As we have shown C is a projective algebra (Theorem 8.16). But here we will give another proof of this fact. Let us consider the following partition E of the algebra C^2 the classes of which are: for any $k \in \omega$

$\|(1, (\neg c)^k)\| = \{(nc, (\neg c)^k) : n \in \omega\} \cup \{((\neg c)^n, (\neg c)^k) : n \in \omega\}$,

$\|(0, kc)\| = \{(nc, kc) : n \in \omega\} \cup \{((\neg c)^n, kc) : n \in \omega\}$.

Notice that this partition is the congruence relation corresponding to the prime filter $\|(1, 1)\| = \{x \in C^2 : (0, 1) \leq x \leq (1, 1)\}$, and $\|(0, 0)\|$ is the prime ideal $\{x \in C^2 : (0, 0) \leq x \leq (1, 0)\}$.

Let us consider the following homomorphisms: $\pi_2 : C^2 \to C$, where $\pi_2((x, y)) = y$, and $\varepsilon : C \to C^2$, where $\varepsilon(kc) = (0, kc)$, $\varepsilon((\neg c)^k) = (1, (\neg c)^k)$ for every $k \in \omega$. Then, it is clear that $\pi_2\varepsilon = Id_C$. From here we conclude that C is projective. □

9.2 Projective Formulas

Let us denote by $\mathcal{P}_m$ a fixed set $x_1, \ldots, x_m$ of propositional variables and by Φ_m the set of all propositional formulas in L_P with variables in $\mathcal{P}_m$. Notice that the m-generated free $MV(C)$-algebra $F_{\mathbf{MV(C)}}(m)$ is isomorphic to $\Phi_m/\equiv$, where $\alpha \equiv \beta$ iff $\vdash (\alpha \leftrightarrow \beta)$ and $\alpha \leftrightarrow \beta = (\alpha \to \beta) \wedge (\beta \to \alpha)$. Subsequently we do not distinguish between the formulas and their equivalence classes. Hence we simply write Φ_m for $F_{\mathbf{MV(C)}}(m)$, and $\mathcal{P}_m$ plays the role of the set of free generators. Since Φ_m is a lattice, we have an order $\leq$ on Φ_m. It follows from the definition of $\to$ that for all $\alpha, \beta \in \Phi_m$, $\alpha \leq \beta$ iff $\vdash (\alpha \to \beta)$.

Let α be a formula of the logic L_P and consider a substitution $\sigma : \mathcal{P}_m \to \Phi_m$ and extend it to all of Φ_m by $\sigma(\alpha(x_1, \ldots, x_m)) = \alpha(\sigma(x_1), \ldots, \sigma(x_m))$. We can consider the substitution as an endomorphism $\sigma : \Phi_m \to \Phi_m$ of the free algebra Φ_m.

Definition 9.9 A formula $\alpha \in \Phi_m$ is called *projective* if there exists a substitution $\sigma : \mathcal{P}_m \to \Phi_m$ such that $\vdash \sigma(\alpha)$ and $\alpha \vdash \beta \leftrightarrow \sigma(\beta)$, for all $\beta \in \Phi_m$.

Notice that the notion of projective formula was introduced for intuitionistic logic in [10].

Observe that we can rewrite any identity $p(x_1, \ldots, x_m) = q(x_1, \ldots, x_m)$ in the variety **MV(C)** into an equivalent one $p(x_1, \ldots, x_m) \leftrightarrow q(x_1, \ldots, x_m) = 1$. So, for **MV(C)** we can replace n identities by one

$$\bigwedge_{i=1}^{n} p_i(x_1, \ldots, x_m) \leftrightarrow q_i(x_1, \ldots, x_m) = 1.$$

Now we are ready to show a close connection between projective formulas and projective subalgebras of the free algebra Φ_m.

Theorem 9.10 *Let A be an m-generated projective subalgebra of the free algebra Φ_m. Then there exists a projective formula α of m variables, such that A is isomorphic to $\Phi_m/[\alpha)$, where $[\alpha)$ is the principal filter generated by $\alpha \in \Phi_m$.*

Proof Suppose A is an m-generated projective subalgebra of Φ_m with generators $a_1, \ldots, a_m$. Then A is a retract of Φ_m, and there exist homomorphisms $\varepsilon : A \to \Phi_m$, $h : \Phi_m \to A$ such that $h\varepsilon = Id_A$, where $\varepsilon(x) = x$ for every $x \in A \subset \Phi_m$. Observe that εh is an endomorphism of Φ_m. We will show now that $\alpha = \bigwedge_{j=1}^{m}(x_j \leftrightarrow \varepsilon h(x_j))$ is a projective formula, namely, that $\vdash \varepsilon h(\alpha)$ and $\alpha \vdash \beta \leftrightarrow \varepsilon h(\beta)$, for all $\beta \in \Phi_m$.

Indeed, $\varepsilon h(\bigwedge_{j=1}^{m}(p_j \leftrightarrow \varepsilon h(p_j))) = \bigwedge_{j=1}^{m}(\varepsilon h(x_j) \leftrightarrow \varepsilon h \varepsilon h(x_j))$, and since $h\varepsilon = Id_A$, we have $\varepsilon h(\bigwedge_{j=1}^{m}(x_j \leftrightarrow \varepsilon h(x_j))) = \bigwedge_{j=1}^{m}(\varepsilon h(x_j) \leftrightarrow \varepsilon h(x_j))$. Thus $\vdash \varepsilon h(\alpha)$. Further, for any $\beta \in \Phi_m$, $\varepsilon h(\beta(x_1, \ldots, x_m)) = \beta(\varepsilon h(x_1), \ldots, \varepsilon h(x_m))$, and since $\alpha \vdash x_j \leftrightarrow \varepsilon h(x_j)$, $j = 1, \ldots, m$, we have $\alpha \vdash \beta \leftrightarrow \varepsilon h(\beta)$.

Since A is an m-generated projective $MV(C)$-algebra, according to the Proposition 9.5, there exist m polynomials $p_1(x_1, \ldots, x_m), \ldots, p_m(x_1, \ldots, x_m)$ such that

$$p_i(x_1, \ldots, x_m) = \varepsilon(a_i) = \varepsilon h(x_i)$$

and

$$p_i(p_1(x_1, \ldots, x_m), \ldots, p_m(x_1, \ldots, x_m)) = p_i(x_1, \ldots, x_m), \ i = 1, \ldots, m.$$

Observe, that $h(x_i) = a_i$. Since the m-generated projective MV-algebra A is finitely presented by the equation $\bigwedge_{j=1}^{m}(x_j \leftrightarrow \varepsilon h(x_j)) = 1$, we have that $A \cong \Phi_m/[\alpha)$. □

Theorem 9.11 *If α is a projective formula of m variables, then $\Phi_m/[\alpha)$ is a projective algebra which is isomorphic to a projective subalgebra of Φ_m.*

Proof Suppose that α is a projective formula of m variables. Then there exists a substitution $\sigma : \mathcal{P}_m \to \Phi_m$ such that $\vdash \sigma(\alpha)$ and $\alpha \vdash \beta \leftrightarrow \sigma(\beta)$, for all $\beta \in \Phi_m$. Since σ is an endomorphism of Φ_m, $\sigma(\Phi_m)$ is a subalgebra of Φ_m. Now we will show that $\sigma(\Phi_m)$ is a retract of Φ_m, i.e. $\sigma^2 = \sigma$. Indeed, since α is a projective formula, $\sigma(\alpha) = 1_{\Phi_m}$, and $\alpha \leq \beta \leftrightarrow \sigma(\beta)$ for all $\beta \in \Phi_m$. But then $\sigma(\alpha) \leq \sigma(\beta) \leftrightarrow \sigma^2(\beta)$, $\sigma(\beta) \leftrightarrow \sigma^2(\beta) = 1_{\Phi_m}$, $\sigma(\beta) = \sigma^2(\beta)$, and $\sigma^2 = \sigma$. Hence $\sigma(\Phi_m)$ is a retract of Φ_m. So, $\sigma(\Phi_m)$ is isomorphic to $\Phi_m/[\alpha)$. □

Thus we have the following correspondence between projective formulas and projective subalgebras of Φ_m. To each m-generated projective subalgebra of m-generated free $MV(C)$-algebra corresponds an m-variable projective formula and to two non-isomorphic m-generated projective subalgebra of m-generated free $MV(C)$-algebra correspond non-equivalent m-variable projective formulas. And two non-equivalent m-variable projective formulas correspond two different m-generated projective subalgebra of m-generated free $MV(C)$-algebra (but they can be isomorphic).

Therefore we arrive at the following

Corollary 9.12 *There exists a one-to-one correspondence between projective formulas with m variables and m-generated projective subalgebras of* Φ_m.

9.3 Unification Problem

Let E be an equational theory. The E-unification problem is: given two terms s, t (built from function symbols and variables), to find a unifier for them, that is, a uniform replacement of the variables occurring in s and t by other terms that makes s and t equal by modulo E. For detail information on unification problem we refer to [10, 11, 19].

Let us be more precise. Let $\mathcal{F}$ be a set of functional symbols and let V be a set of variables. Let $T_{\mathcal{F}}(V)$ be the term algebra built from $\mathcal{F}$ and V, and $T_{\mathcal{F}_m}(V)$ be the term algebra of m-variable terms. Let E be a set of identities of type $p(x_1, \ldots, x_m) = q(x_1, \ldots, x_m)$, where $p, q \in T_{\mathcal{F}_m}(V)$.

Let $\mathbf{V}$ be the variety of algebras over $\mathcal{F}$ axiomatized by the identities from E.

A unification problem modulo E is a finite set of pairs

$$\mathcal{E} = \{(s_j, t_j) : s_j, t_j \in T_{\mathcal{F}_m}(V), j \in J\}$$

for some finite set J. *A solution to* (or *a unifier for*) $\mathcal{E}$ is a substitution (or an endomorphism of the term algebra $T_{\mathcal{F}_m}(V)$) σ (which is extension of the map $s : V_m \to T_{\mathcal{F}_m}(V)$, where V_m ($= \{x_1, \ldots, x_m\}$) is the set of m variables) such that the identity $\sigma(s_j) = \sigma(t_j)$ holds in every algebra of the variety $\mathbf{V}$. The problem $\mathcal{E}$ is *solvable* (or *unifiable*) if it admits at least one unifier.

Let $(X, \preceq)$ be a quasi-ordered set (i.e. $\preceq$ is a reflexive and transitive relation). A μ-set [11] for $(X, \preceq)$ is a subset $M \subseteq X$ such that: (1) every $x \in X$ is less or equal to some $m \in M$; (2) all elements of M are mutually $\preceq$-incomparable. There might be no μ-set for $(X, \preceq)$ (in this case we say that $(X, \preceq)$ has *type* 0) or there might be many of them, due to the lack of antisymmetry. However all μ-sets for $(X, \preceq)$, if any, must have the same cardinality. We say that $(X, \preceq)$ has *type* $1, \omega, \infty$ iff it has a μ-set of cardinality 1, of finite (greater than 1) cardinality or of infinite cardinality, respectively.

Substitutions are compared by instantiation in the following way: we say that $\sigma : T_{\mathcal{F}_m}(V) \to T_{\mathcal{F}_m}(V)$ is *more general than* $\tau : T_{\mathcal{F}_m}(V) \to T_{\mathcal{F}_m}(V)$ (written as $\tau \preceq \sigma$) iff there is a substitution $\eta : T_{\mathcal{F}_m}(V) \to T_{\mathcal{F}_m}(V)$ such that for all $x \in V_m$ we have $E \vdash \eta(\sigma(x)) = \tau(x)$. The relation $\preceq$ is quasi-order.

Let $U_E(\mathcal{E})$ be the set of unifiers for the unification problem $\mathcal{E}$; then $(U_E(\mathcal{E}), \preceq)$ is a quasi-ordered set.

We say that an equational theory E has:

1. Unification type 1 iff for every solvable unification problem $\mathcal{E}$, $U_E(\mathcal{E})$ has type 1;
2. Unification type ω iff for every solvable unification problem $\mathcal{E}$, $U_E(\mathcal{E})$ has type ω;
3. Unification type ∞ iff for every solvable unification problem $\mathcal{E}$, $U_E(\mathcal{E})$ has type 1 or ω or ∞—and there is a solvable unification problem $\mathcal{E}$ such that $U_E(\mathcal{E})$ has type ∞;
4. Unification type nullary, if none of the preceding cases applies.

Following Ghilardi [10], who has introduced the relevant definitions for E-unification from an algebraic point of view, by an algebraic unification problem we mean a finitely presented algebra A of $\mathbf{V}$. In this context an E-unification problem is simply a finitely presented algebra A, and a solution for it (also called a unifier for A) is a pair given by a projective algebra P and a homomorphism $u : A \to P$. The set of unifiers for A is denoted by $U_E(A)$. A is said to be *unifiable* or solvable iff $U_E(A)$ is not empty. Given another algebraic unifier $w : A \to Q$, we say that u is more general than w, written $w \preceq u$, if there is a homomorphism $g : P \to Q$ such that $w = gu$.

The set of all algebraic unifiers $U_E(A)$ of a finitely presented algebra A forms a quasi-ordered set with the quasi-ordering $\preceq$.

The algebraic unification type of an algebraically unifiable finitely presented algebra A in the variety $\mathbf{V}$ is now defined exactly as in the symbolic case, using the quasi-ordering set $(U_E(A), \preceq)$. If m-generated finitely presented algebra of an equational class $\mathbf{V}$ is projective, then Id_A will be most general unifier for A.

Theorem 9.13 *The unification type of the equational class* $\mathbf{MV}(\mathbf{C})$ *is* 1, *i.e. unitary.*

Proof The proof of the theorem immediately follows from Theorem 8.23. □

9.4 Structural Completeness

A logic L is structurally complete if every rule that is admissible (preserves the set of theorems) should also be derivable. In a logic, a rule of inference is admissible in a formal system if the set of theorems of the system does not change when that rule is added to the existing rules of the system.

A Tarski-style consequence relation is a relation $\vdash$ between sets of formulas, and formulas, such that

- $\alpha \vdash \alpha$,
- if $\Gamma \vdash \alpha$, then $\Gamma, \Delta \vdash \alpha$.

A consequence relation such that if $\Gamma \vdash \alpha$, then $\sigma(\Gamma) \vdash \sigma(\alpha)$ for all substitutions σ is called structural.

More precisely. If L is a logic, an *L-unifier* of a formula φ is a substitution σ such that $\vdash_L \sigma(\varphi)$. A formula which has an L-unifier is called *L-unifiable*. An inference rule is an expression of the form Γ/φ, where φ is a formula, and Γ is a finite set of formulas. An inference rule Γ/φ is *derivable* in a logic L, if $\Gamma \vdash_L \varphi$. The rule $\Gamma \vdash_L \varphi$ is *L-admissible*, if every common L-unifier of Γ is also an L-unifier of φ.

We can identify propositional formulas with terms in the language of MV-algebras in a natural way. A *valuation* in an MV-algebra A is a homomorphism v from the term algebra to A. If φ is a k-variable formula, $(a_1, \ldots, a_k) \in A^k$, and v is the assignment such that $v(p_i) = a_i$, we also write $\varphi(a_1, \ldots, a_k) = v(\varphi)$. A valuation v satisfies a formula φ if $v(\varphi) = 1$, and it satisfies a rule Γ/φ if $v(\varphi) \neq 1$ for some $\alpha \in \Gamma$, or $v(\varphi) = 1$. A rule Γ/φ is *valid* in an MV-algebra A, written as $A \models \Gamma/\varphi$, if the rule is satisfied by every valuation in A. In other words, $A \models \Gamma/\varphi$ if and only if the open first-order formula

$$\bigwedge_{\alpha \in \Gamma} (\alpha = 1) \Rightarrow \varphi = 1$$

is valid in A. Conversely, validity of open formulas (or equivalently, universal sentences) in A can be reduced to validity of rules. Any open formula Φ can be expressed in the conjunctive normal form as $\Phi = \bigwedge_{i<k} \Phi_i$, where each Φ_i is a clause: a disjunction of atomic formulas (i.e., equations) and their negations. Then $A \models \Phi$ iff $A \models \Phi_i$ for each $i < k$, and a clause

$$\bigvee_{i<n} (\varphi_i = \psi_i) \vee \bigvee_{i<m} (\varphi_i' \neq \psi_i')$$

is valid in A iff validates the rule

$$\{\varphi_i' \leftrightarrow \psi_i' \mid i < m\} / \{\varphi_i \leftrightarrow \psi_i \mid i < n\}.$$

Łukasiewicz logic Ł is algebraizable, and the variety of MV-algebras is its equivalent algebraic semantics, using the translation between propositional formulas and identities described above. We thus have (cf. [20]):

Claim 9.14 [21] *A rule Γ/φ is valid in all MV-algebras if and only if it is derivable in Ł.*

As another corollary to algebraizability of Ł, free MV-algebras can be described as Lindenbaum algebras of Ł: the Lindenbaum algebra consists of equivalence classes of formulas using elements of generators X as propositional variables modulo the equivalence relation $\varphi \sim \psi$ iff $\vdash_{Ł} \varphi \leftrightarrow \psi$, with operations defined in the natural way. Note that valuations in this Lindenbaum algebra correspond to substitutions whose range consists of formulas using variables from X, and a formula φ is satisfied under a valuation given by such a substitution σ if and only if $\vdash_{Ł} \sigma(\varphi)$. We obtain the following characterization of admissibility:

Claim 9.15 [21] *For any rule Γ/φ, the following are equivalent:*

(i) Γ/φ is admissible.
(ii) Γ/φ is valid in all free MV-algebras.
(iii) Γ/φ is valid in all free MV-algebras over finite sets of generators.

Let us note that we will have the same assertions if we change the Łukasiewicz logic Ł with logic L_P. Then we can reformulate the Claim 9.15 in the following way:

The logic L_P is structurally complete iff the variety **MV(C)** *coincides with the quasi variety generated by all free $MV(C)$-algebras over finite sets of generators.*

Let us formulate the following property for a logic L:

(SC) $\quad \alpha \vdash \beta \in T \quad \Leftrightarrow (\forall \varphi : Form(\mathfrak{L}) \to Form(\mathfrak{L}))[\varphi(\alpha) \in T \Rightarrow \varphi(\beta) \in T]$,

where T is the set of all theorems of the logic L, φ is an endomorphism of the algebra $(F; \to, \neg, 0, 1)$ which is a free algebra in the class of algebras of the type (2,1,0,0). Let us note that this condition is equivalent to the notion of a structural completeness [22] in the sense of Pogorzelski, i.e. any structural admissible rule of a logic is derivable.

(SCL) $\quad \alpha^n \to \beta \in T$, for some positive integer n, $\Leftrightarrow (\forall \varphi : F \to F)[\varphi(\alpha) \in T \Rightarrow \varphi(\beta) \in T]$,

where T is the set of all theorems of the logic L, φ is an endomorphism of the algebra $(F; \to, \neg, 0, 1)$ which is a free algebra in the class of algebras of the type (2,1,0,0). Let us note that, since according to deduction theorem in Łukasiewicz logic: $\alpha \vdash \beta$ if and only if $\vdash \alpha^n \to \beta$ for some positive integer n, the property is equivalent to the notion of a structural completeness in the sense of Pogorzelski, i.e. any structural admissible rule of a logic is derivable.

In algebraic terms the property has the following formulation:

- $\alpha^n \to \beta = 1$, for some positive integer $n \Leftrightarrow (\forall\varphi : Form(\mathfrak{L}) \to Form(\mathfrak{L}))[\varphi(\alpha) = 1 \Rightarrow \varphi(\beta) = 1]$,

where φ is an endomorphism of the ω-generated free algebra $(F; \to, \neg, 0, 1)$ in the variety of MV-algebras.

Recall that L_P is a logic corresponding to variety **MV(C)**, i.e. L_P is the extension of Łukasiewicz logic by the Lukasievicz formula $\neg((\neg\alpha \to \alpha) \to \neg(\neg\alpha \to \alpha)) \leftrightarrow ((\alpha \to \neg\alpha) \to \neg(\alpha \to \neg\alpha))$, the theorems of which coincides with formulas that is valid in all $MV(C)$-algebras.

Theorem 9.16 *The logic L_P is structurally complete.*

Proof Let us suppose that $\alpha \to \beta$ is m variable term. It is evident that if $\alpha^n \to \beta = 1$, then $(\forall\varphi : F \to F)[\varphi(\alpha) = 1 \Rightarrow \varphi(\beta) = 1]$.

Now suppose that $\alpha^n \to \beta \neq 1$ for all positive integers n and $\varphi : F \to F$ is an endomorphism such that $\varphi(\alpha) = 1$. Therefore, there exists m generators of $MV(C)$-algebra C where $\alpha > \beta$ on the generators $a_1, \ldots, a_m \in C$, i.e. $\alpha(a_1, \ldots, a_m) > \beta(a_1, \ldots, a_m)$ and $\alpha(a_1, \ldots, a_m)$ belongs to a prime filter, say J, and, since $\alpha^n(a_1, \ldots, a_m) > \beta(a_1, \ldots, a_m)$ for all positive integers n, $\beta(a_1, \ldots, a_m)$ does not belong to J. Observe that J is either the minimal prime filter $\{1\}$ or maximal filter $\{(\neg c)^k : k \in \omega\}$. Then, C/J is a chain $MV(C)$-algebra such that $\alpha(a_1/J, \ldots, a_m/J) = 1$ and $\beta(a_1/J, \ldots, a_m/J) \neq 1$. According to Theorem 3.7, C/J is projective, which is either two-element Boolean algebra or $MV(C)$-algebra C. Hence, there exist homomorphisms $h : F(m) \to C/J$ and $\varepsilon : C/J \to F(m)$ such that $h\varepsilon = Id_{C/J}$. Then $\varepsilon h : F(m) \to F(m)$ is an endomorphism such that $\varepsilon h(\alpha) = 1$ and $\varepsilon h(\beta) \neq 1$.

Now we give another proof of this theorem. We show that the variety **MV(C)** coincides with the quasi variety generated by all free $MV(C)$-algebras over finite sets of generators. Indeed, since C is projective, C is a subalgebra of a free $MV(C)$-algebras over finite sets of generators. But quasi variety $\mathcal{QV}(C)$ generated by C coincides with the variety $\mathcal{V}(C)$. □

Corollary 9.17 *Among the extensions of Łukasiewicz logics only classical logic and the logic L_P are structurally complete.*

Proof Let L_0 be a logic distinct from classical logic and the logic L_P. The rule $(3(p \wedge \neg p))^2/p$ is admissible. Indeed, there is no substitution σ such that $\vdash_{L_0} (3(\sigma(p) \wedge \neg\sigma(p)))^2$. Only in the case when $\sigma(p)$ has the value t, such that $t \leq 1/2$ and $2t \geq 1/2$, the valuation of $(3(\sigma(p) \wedge \neg\sigma(p)))^2$ has the value 1. But there is no formula which is equivalent to constant t, since we have no constant t. So, the rule $(3(p \wedge \neg p))^2/p$ is admissible. But $(3(p \wedge \neg p))^2 \to p$ is not a theorem of L_0, because it is not logically true. At the same time the rule is derivable in classical logic and the logic L_P. □

Let us notice that the result of Corollary 9.17 was obtained by J. Gispert in [23]. Let us note that structural completeness for the logic of perfect algebras L_P was announced in [24–26].

We also mention related works on structural completeness and admissibility in MV-algebras/Łukasiewicz logic [21, 27–29].

References

1. Di Nola, A., Lettieri, A.: Perfect MV-algebras are categorically equivalent to abelian ℓ-groups. Studia Logica **88**, 467–490 (1994)
2. Belluce, L.P., Di Nola, A., Gerla, B.: Perfect MV-algebras and their logic. Appl. Categ. Struct. **15**(1–2), 135–151 (2007)
3. Cignoli, R., Torens, A.: Free algebras in varieties of Glivenko MTL-algebras satisfying the equation $2(x^2) = (2x)^2$. Studia Logica **83**, 157–181 (2006)
4. Scarpellini, B.: Die Nichtaxiomatisierbarkeit des Unendlichwertigen Pradikatenkalkulus von Łukasiewicz. J. Symbol. Logic **27**, 159–170 (1962)
5. Belluce, L.P., Chang, C.C.: A weak completeness theorem for infinite valued predicate logic. J. Symb. Logic **28**, 43–50 (1963)
6. Belluce, L.P., Di Nola, A.: The MV-algebra of first order Łukasiewicz logic. Tatra Mt. Math. Publ. **27**(1–2), 7–22 (2007)
7. Mundici, D.: Interpretation of AF C^*-algebras in Łukasiewicz sentential calculus. J. Funct. Anal. **65**, 15–63 (1986)
8. Mundici, D.: Turing complexity of the Behncke-Leptin C^*-algebras with a two-point dual. Ann. Math. Artif. Intell. **26**, 287–294 (1992)
9. Lorenzen, P.: Einfuhrung in die Operative Logik und Mathematik. Grundlehren der Mathematischen Wissenschaften, vol. 78, Springer (1955)
10. Ghilardi, S.: Unification in intuitionistic and De Morgan logic. J. Symb. Logic 859–880 (1999)
11. Ghilardi, S.: Best solving modal equations. Ann. Pure Appl. Logic **102**(3), 183–198 (2000)
12. Ghilardi, S.: Unification, finite duality and projectivity in varieties of heyting algebras. APAL **127**, 99–115 (2004)
13. Di Nola, A, Grigolia, R.: Projective MV-algebras and their automorphism groups. J. Multi-Valued Logic Soft Comput. **9**, 291–317
14. McKenzie, R.: An Algebraic Version of Categorical Equivalence for Varieties and More General Algebraic Categories. In: Logic and Algebra (Pontignano, 1994), vol. 180, Lecture Notes in Pure and Applied Mathematics, pp. 211–243. Dekker, New York (1996)
15. Mundici, D.: Advanced Łukasiewicz calculus and MV-algebras. Trends in Logic, vol. 35. Springer, New York (2011)
16. Baker, K.A.: Free vector lattices. Can. J. Math. **20**, 58–66 (1968)
17. Beynon, W.M.: Combinatorial aspects of piecewise linear maps. J. Lond. Math. Soc. **31**(2), 719–727 (1974)
18. Mundici, D.: The Haar theorem for lattice-ordered abelian groups with order-unit. Discrete Contin. Dyn. Syst. **21**, 537–549 (2008)
19. Ghilardi, S.: Unification through projectivity. J. Logic Comput. **7**, 733–752 (1997)
20. Horn, A.: Logic with truth values in a linearly ordered Heyting algebra. J. Symbol. Logic **34**, 395–408 (1969)
21. Jerabek, E.: Admissible rules of Łukasiewicz logic. J. Logic Comput. **20**, 425–447 (2010)
22. Pogorzelski, W.A.: Structural completeness of the propositional calculus. Bull. Acad. Polon. Sci. Ser. Math. Astr. Phys. **19**, 349–351 (1971)
23. Gispert, J.: Least V-quasivarieties of MV-algebras. Fuzzy Sets Syst. (2014). doi:10.1016/j.fss.2014.07.011

24. Di Nola, A., Grigolia, R., Spada, L.: On the Logic of Perfect MV-algebras: Projectivity. Unification, Structurally Completeness. Research Workshop in Duality Theory in Algebra, Logic and Computer Science Workshop II, Oxford (2012)
25. Di Nola, A., Grigolia, R., Lenzi, G.: On the logic of perfect MV-algebras. Logic, Algebra and Truth Degrees (LATD2014), Austria, Vienna (2014)
26. Gispert, J.: Quasi varieties of MV-algebras and structurally complete Łukasiewicz logics. Logic, Algebra and Truth Degrees (LATD2014), Austria, Vienna (2014)
27. Cintula, P., Metcalfe, G.: Structural completeness in fuzzy logics. Notre Dame J. Formal Logic **50**(2), 153–183 (2009)
28. Jerabek, E.: Bases of admissible rules of Łukasiewicz logic. J. Logic Comput. **20**, 1149–1163 (2010)
29. Wojtylak, P.: On structural completeness of many-valued logics. Studia Logica **37**, 139–147 (1978)

Chapter 10
The Logic of Quasi True

10.1 Introduction

We introduce a new logic CL, which is an extension of the infinitely valued Łukasiewicz logic Ł, the language of which enriched by 0-ary connective $\mathbf{c}$ that is interpreted as *quasi false*, the algebraic counterpart of which are algebras from a quasi variety of the variety generated by the perfect MV-algebras. For this aim we introduce a new class **CL** of algebras which is a quasi variety and the algebras from this quasi variety we name CL-algebras. Adding a new inference rule to the logic CL, thereby increased a deducibility power, we introduce the logic CL^+ and defining the notion of quasi true (q-true) formulas it is proved the completeness theorem for this logic.

10.2 CL-Algebras

A CL-algebra $A = (A, 0, \mathbf{c}, \neg, \oplus)$ is an abelian monoid $(A, 0, \oplus)$ equipped with a constant element $\mathbf{c}$ and a unary operation $\neg$ such that $\neg\neg x = x$, $x \oplus \neg 0 = \neg 0$, and $y \oplus \neg(y \oplus \neg x) = x \oplus \neg(x \oplus y)$. We set $1 = \neg 0$ and $x \odot y = \neg(\neg x \oplus \neg y)$ [1].

The above assures that $A = (A, 0, \neg, \oplus)$ is an MV-algebra. Additionally,

(i) $2(x^2) = (2x)^2$,
(ii) $2\mathbf{c} \odot \neg\mathbf{c} = \mathbf{c}$,
(iii) $\mathbf{c} \odot (\neg x \vee x) \wedge (x \wedge \neg x) = 0$,
(iv) $\mathbf{c} \to \neg\mathbf{c} = 1$,
(v) $x \vee \neg\mathbf{c} = 1 \Rightarrow x = 1$.

Hereinafter we denote CL-algebra as $(A, \mathbf{c})$, where A is an MV-algebra.

Comment. The identity (1) says that a CL-algebra is a member of the subvariety $V(C)$ (= the variety generated by all perfect MV-algebras) of the variety of all

A. Di Nola et al., *Fuzzy Logic of Quasi-Truth: An Algebraic Treatment*,
Studies in Fuzziness and Soft Computing 338, DOI 10.1007/978-3-319-30406-9_10

MV-algebras. The second (2) says that $\mathbf{c} \neq \mathbf{c} \vee \neg\mathbf{c}$ and $\mathbf{c} \neq 1$. (3) Says that $\mathbf{c}$ is the atom in a totally ordered CL-algebra. (4) Says that $\mathbf{c} \leq \neg\mathbf{c}$. (5) Says that $\mathbf{c} \neq 0$ and exclude the MV-algebra with constant $(\mathbf{2} \times C, (0, c))$ (where $\mathbf{2}$ is two-element Boolean algebra), since the quasi identity $x \vee \neg\mathbf{c} = 1 \Rightarrow x = 1$ does not hold when $x = (0, 1)$; indeed $(0, 1) \vee \neg(0, c) = (1, 1)$ and $(0, 1) \neq (1, 1)$. Denote the class of all CL-algebras by **CL**. We assume that **CL** contains one-element CL-algebra. It is obvious the following

Theorem 10.1 *The class* **CL** *is a quasivariety.*

Lemma 10.2 *Let $(A, \mathbf{c})$ be a totally ordered CL-algebra. There is no element $x \in A$ such that $n\mathbf{c} < x < (n+1)\mathbf{c}$ for some $n \in Z^+$.*

Proof Let us suppose that there exists an element $x \in A$ such that $n\mathbf{c} < x < (n+1)\mathbf{c}$ for some $n \in Z^+$. Then $0 < (\neg\mathbf{c})^n \odot x < c$. Notice, that $0 < (\neg\mathbf{c})^n \odot x$ since if $0 = (\neg\mathbf{c})^n \odot x$, then $n\mathbf{c} \geq x$ which contradicts to the initial condition $n\mathbf{c} < x$. But this contradicts to the condition that $\mathbf{c}$ is the atom in a totally ordered CL-algebra. □

Corollary 10.3 *The CL-algebra $(C, \mathbf{c})$ is a subalgebra of every CL-algebra $(A, \mathbf{c})$.*

Let A be an MV-algebra and $P \in \mathcal{F}(A)$ where $\mathcal{F}(A)$ is the set of all prime filters of A. We say that P is a *Chang's filter* iff A/P is isomorphic to C, where C is Chang's algebra. We say that F is a CL-filter if it is an intersection of Chang's filters. From here we conclude that $\{1\}$ is CL-filter. We can characterize Chang's filters as follows

Lemma 10.4 *Let $(A, \mathbf{c})$ be an CL-algebra and $P \in \mathcal{F}(A)$, then the following conditions are equivalent:*

(1) P is a Chang's filter,
(2) P does not contain $\neg\mathbf{c}$ and P is maximal filter with this condition ($\neg\mathbf{c} \notin P$).

Proof (1) $\Rightarrow$ (2). Let $P \in \mathcal{F}(A)$ be a Chang's filter of a CL-algebra A. Then A/P is isomorphic to C. Let us suppose that P' is a CL-filter such that $P \subset P'$ and $P \neq P'$. Since P' is a Chang's filter, we have that $P' = \bigcap_{i \in I} F_i$ where F_i is a Chang's filter for every $i \in I$. Then A/P is a homomorphic image of A/F_i for some $i \in I$. It means that A/P is not isomorphic to C which contradicts to the assumption. Any Chang's filter P does not contain $\neg\mathbf{c}$ and $(\neg\mathbf{c})^n$ as well. Indeed, if $(\neg\mathbf{c})^n \in P$, then $(\neg\mathbf{c})^n \cong_P (\neg\mathbf{c})^m$ for any $n, m \in Z^+$ that is impossible. (2) $\Rightarrow$ (1). Let us suppose that P does not contain $\neg\mathbf{c}$ and P is maximal filter with this condition ($\neg\mathbf{c} \notin P$). Then $(\neg\mathbf{c})^n \ncong_P (\neg\mathbf{c})^m$ for any $n, m \in Z^+$ such that $n > m$ (or $m > n$). Indeed, if $(\neg\mathbf{c})^n \cong_P (\neg\mathbf{c})^m$, then $(\neg\mathbf{c})^n \to (\neg\mathbf{c})^m = nc \oplus (\neg\mathbf{c})^m = (\neg\mathbf{c})^{n-m} \in P$. But it is impossible. Taking into account that A/P is totally ordered, according to axioms (3) and (5) we have that $A/P \cong (C, c)$. □

Lemma 10.5 *Let $(A, \mathbf{c})$ be CL-algebra. If $a \in A$ and $a \nleq (\neg\mathbf{c})^n$ for any $n \in Z^+$, then there exists Chang's filter P of A such that $a \notin P$.*

Proof Let us consider a CL-filter of A which is maximal with respect to the property that $a \notin F$. We show that F is a Chang's filter. Let $x, y \in A$ and assume that $x \vee y \in F$ and $x, y \notin F$. Thus the CL-filter generated by F and the element x would contain the element a, i.e. $a \geq fx^p$ for some $f \in F$ and $p \in Z^+$. Similarly, the filter generated by F and y would also contain a, i.e., $a \geq f'y^q$ for some $f' \in F$ and $q \in Z^+$. Let $n = max(p, q)$. Then clearly $ff' \in F$ and from the above we have $a \geq ff' \odot x^n$ and $a \geq ff' \odot y^n$. From here we get $a \geq (ff' \odot x^n) \vee (ff' \odot y^n) = ff' \odot (x^n \vee y^n)$. Thus $a \geq ff'$ which implies the conradiction that $a \in F$. Since F is a maximal CL-filter, we have that F is Chang's filter. □

From this lemma we immediately have

Corollary 10.6 *The intersection of all Chang's filters of* $(A, \mathbf{c})$ *is equal to* $\{1\}$.

Let $(A, \mathbf{c})$ be a CL-algebra which is a product of copies of the algebra $(C, \mathbf{c})$ and $B(A)$ the Boolean skeleton of the MV-algebra A. Let M be a filter of $B(A)$. F is a CL-filter of $(A, \mathbf{c})$ if it is an MV-filter of A and $F = [M)$ where M is a Boolean filter of $B(A)$, i.e. the filter F is generated by some Boolean filter M of $B(A)$ and denote this filter by $F(M)$. From this definition we have that $\{1\}$ is a CL-filter. So, we conclude that a maximal CL-filter F of $(A, \mathbf{c})$ is generated by a maximal Boolean filter of $F \cap B(A)$. Let $(A_1, \mathbf{c})$ be a CL-subalgebra of the algebra $(A, \mathbf{c})$. Then the intersection $F \cap A_1$ of a prime CL-filter of $(A, \mathbf{c})$ with the subalgebra A_1 will be also a prime (and maximal as well) CL-filter of the algebra $(A_1, \mathbf{c})$. So, the factor algebra $(A, \mathbf{c})/F$ by a Chang's filter will be subdirectly irredusible which will be totally ordered CL-algebra that is isomorphic to $(C, \mathbf{c})$. Therefore

Theorem 10.7 *Any* CL*-algebra* $(A, \mathbf{c})$ *is represented as a subdirect product of* $(A, \mathbf{c})/F_i$, $i \in I$, *where* F_i *is a Chang's filter of* $(A, \mathbf{c})$ *and* $(A, \mathbf{c})/F_i \cong (C, \mathbf{c})$.

As in the variety of ℓ-groups we can define the *polar* of a subset $M \subset A$ of a CL-algebra $(A, \mathbf{c})$ as the set $M^\perp = \{a \in A : \forall x \in M \; x \wedge a = 0\}$.

Theorem 10.8 *The polar of* $\{c\}^\perp = \{0\}$ *for any* CL*-algebra* $(A, \mathbf{c})$.

Proof Assume that for a non-zero element $x \in A$, we have that $c \wedge x = 0$. Then $\neg c \vee \neg x = 1$. By Axiom 6 we have that $\neg x = 1$. That is $x = 0$, a contradiction. □

We express next property by the following

Theorem 10.9 *Let us suppose that* A *is a* $MV(C)$*-algebra and* $(A, \mathbf{c})$ *is a* CL*-extension of* A*. Such kind extension is unique.*

Proof Let us assume that (A, z) is a CL-algebra too. Then by Axiom 3 we get for $x = z$, $(c \odot \neg z) \wedge z = 0$. Checking the equality over a CL-chain, since z is non-zero we have: $(c \odot \neg z) = 0$. This implies that $c \leq z$. Symmetrically, we also have $z \leq c$. Hence $c = z$. □

Theorem 10.10 *The quasi variety* **CL** *is generated by the algebra* $(C, \mathbf{c})$. *Moreover,* $\mathbf{CL} = \mathbf{SP}(C, \mathbf{c})$, *where* **S** *is the operator of taking a subalgebra and* **P** *is the operator of taking a direct product.*

Proof It is clear that **CL** is an axiomatized class. So, **CL** is **SP(CL)**. But every algebra from **CL** is a subdirect product of algebras that are isomorphic to $(C, \mathbf{c})$. Hence, $\mathbf{CL} = \mathbf{SP}(C, \mathbf{c})$. □

10.3 Logics CL and CL^+

In this section we define logic CL, the algebraic counterparts of which are CL-algebras. The language of the logic CL consists of the propositional variables $p_1, p_2, p_3, \ldots$, propositional constant c, logical connectives $\to, \neg$. The formulas are defined as usual. The following formulas are axioms:

L1. $\alpha \to (\beta \to \alpha)$,
L2. $(\alpha \to \beta) \to ((\beta \to \gamma) \to (\alpha \to \gamma))$,
L3. $(\neg\alpha \to \neg\beta) \to (\beta \to \alpha)$,
L4. $((\alpha \to \beta) \to \beta) \to ((\beta \to \alpha) \to \alpha)$,
Lp. $2(\alpha^2) \leftrightarrow (2\alpha)^2$,
CL. $c \to \neg c$,
CL1. $2c \odot \neg c \leftrightarrow c$,
CL2. $(c \to (\neg\alpha \wedge \alpha)) \vee (\alpha \vee \neg\alpha)$.

Inference rules: MP. $\alpha, \alpha \to \beta \Rightarrow \beta$, R1. $\alpha \vee \neg c \Rightarrow \alpha$.

We say that α is q-true (or q-tautology) iff $\neg\alpha \to \alpha$ is a 1-true (or 1-tautology). Semantically, we say that α is q-true if $e(\alpha) \in \neg RadC$ for every evaluation $e : Var \cup \{c\} \to (C, c)$. It is hold the following

Theorem 10.11 $\vdash_{CL} 2(\neg c)^n$ *for any* $n \in Z^+$.

Proof From the axiom CL we have $c \to \neg c$ is a theorem of CL. But $(c \to \neg c) \equiv (\neg c \oplus \neg c)$. According to the Axiom Lp we have $\vdash_{CL} 2((\neg c)^2))^2) \leftrightarrow (2(\neg c)^2)^2$. So, $\vdash_{CL} 2(\neg c)^4$ and so on by induction. □

Theorem 10.12 (Completeness theorem) α *is* 1*-true iff* $\vdash_{CL} \alpha$.

Proof Notice, that any axiom of the logic CL is 1-tautology and the inference rules preserves 1-tautology. So, if $\vdash_{CL} \alpha$, then α is a 1-tautology. Now suppose that α is not theorem of CL. Then $[\alpha] \neq 1$ in the Lindenbaum algebra $\mathfrak{L}$ of the logic CL. As we know $\mathfrak{L}$ is a subdirect product of the copies of $(C, \mathbf{c})$. Then in one of the factors $(C, \mathbf{c})$ for some projection $\pi_i : \mathfrak{L} \to (C, \mathbf{c})$ we have $\pi_i([\alpha]) \neq 1$. So, a is not 1-tautology. □

Theorem 10.13 *If* α *is* q*-true, then* $\vdash_{CL} \neg\alpha \to \alpha$.

Proof Let us suppose that α is q-true. It means that $e(\alpha) \in \neg RadC$ for any evaluation $e : V \cup \{c\} \to (C, \mathbf{c})$. Therefore $2e(\alpha) = \neg e(\alpha) \to e(\alpha) = 1$ for any evaluation e. It means that $\neg\alpha \to \alpha$ is 1-tautology. So, according to completeness theorem, $\vdash_{CL} \neg\alpha \to \alpha$. □

Corollary 10.14 *If α is q-true, then $\vdash_{CL} (\neg\alpha \vee \alpha) \to \alpha$.*

Proof Let us suppose that α is q-true. According to the Theorem 10.10 $\neg e(\alpha) \leq e(\alpha)$. Therefore, $e(\neg\alpha) \vee e(\alpha) = e(\alpha)$. So, $(\neg\alpha \vee \alpha) \to \alpha$ is 1-tautology and, hence, according to the Theorem 10.9, $\vdash_{CL} (\neg\alpha \vee \alpha) \to \alpha$. □

Now let us add to the inference rule of the logic CL the following rule: R2. $(\alpha \vee \neg\alpha) \to \alpha \Rightarrow \alpha$ and denote this new logic by CL^+. As we see all q-true formulas are deducible in CL^+. So, we have

Theorem 10.15 (Completeness) *If α is q-true, then $\vdash_{CL^+} \alpha$.*

Proof If α is q-true, then $\vdash_{CL} (\neg\alpha \vee \alpha) \to \alpha$ (Corollary 10.6). Then by R4 $\vdash_{CL^+} \alpha$. □

Theorem 10.16 (Soundness) *If $\vdash_{CL^+} \alpha$, then α is q-true.*

Proof It is routine to check that any axiom of CL^+ is 1-tautology and, hence, q-true, and if any antecedent of any inference rule of CL^+ is q-true, then the consequent is q-true. So, $\vdash_{CL^+} \alpha$ implies α is q-true. □

Now we analyse what kind of balance exists between classical logic Cl and the logic CL^+. For this aim for every formula α of the logic CL^+ define its translation $tr(\alpha)$ into classical logic Cl as follows: (1) if α is a propositional variable p, then $tr(\alpha) = \alpha$; (2) $tr(c) = p \wedge \neg p$; (3) $tr(\alpha \to \beta) = tr(\alpha) \to tr(\beta)$; (4) $tr(\neg\alpha) = \neg tr(\alpha)$. It holds

Theorem 10.17 $\vdash_{CL^+} \alpha$ *iff* $\vdash_{Cl} tr(\alpha)$.

Proof It is obvious that if $\vdash_{CL^+} \alpha$, then α is q-true. Therefore $tr(\alpha)$ will be classical tautology and, hence, $\vdash_{Cl} tr(\alpha)$. If α is not a theorem of CL^+, then α is not q-true. Therefore $tr(\alpha)$ will not be classical tautology, and, hence, will not be a theorem of classical logic Cl. □

It is easy to prove the following

Theorem 10.18 *If $\vdash_{Cl} \alpha$, then $\vdash_{CL^+} \alpha$.*

Reference

1. Chang, C.C.: Algebraic analysis of many-valued logics. Trans. Am. Math. Soc. **88**, 467–490 (1958)

Chapter 11
Perfect Pavelka Logic

11.1 Introduction

A conventional approach in mathematical propositional logic is, after defining a formal language i.e. atomic formulas, logical connectives and the set of well-formed formulas, to interpret semantically these formulas in a suitable algebraic structures. This applies both to classical two valued logic and more general logics, e.g. Łukasiewicz logic as we seen in previous chapters. In classical logic these algebraic structures are Boolean algebras, in Hájek's Basic Fuzzy Logic [1], for example, the suitable structures are BL-algebras and in Łukasiewicz logic MV-algebras. *Tautologies* of a logic are those formulas that obtain the top value **1** in all interpretations in all suitable algebraic structures; for this reason tautologies are sometimes called **1**-tautologies to distinguish them from possible weaker notions of tautologies in fuzzy logics (a detailed treatment of such alternatives can be found in [2]). For example, tautologies in Basic Fuzzy Logic are exactly the formulas that obtain value **1** in all interpretations in all BL-algebras. The next step is to fix the axiom schemata and the rules of inference: a well-formed formula is a *theorem* if it is either an axiom or obtained recursively from axioms by using rules of inference finitely many times. *Completeness* of the logic means that tautologies and theorems coincide; classical sentential logic, Basic Fuzzy sentential logic and Łukasiewicz sentential logic are complete logics.

Many-valued logic can be understood also as a logic of *partially* provable or *partially* true formulas. This is what Jan Pavelka inteded in his three seminal papers Fuzzy Sentential Logic I, II, III [3–5]. Indeed, Pavelka intended to provide solid grounds to Fuzzy Logic, understood as a particular many-valued logic. This meant a generalization of classical logic in such a way that axioms, theories, theorems, and tautologies need not be only fully true or fully false, but may be also true to a degree and, therefore, giving rise to such concepts as fuzzy theories, fuzzy set of axioms, many-valued rules of inference, provability degree, truth degree, fuzzy consequence operation, etc. Pavelka was inspired by paper [6], where Goguen argued that the algebraic structure of Fuzzy Logic should be a (complete) residuated lattice in the same sense as Boolean algebra is the algebraic counterpart of Boolean Logic. Pavelka

A. Di Nola et al., *Fuzzy Logic of Quasi-Truth: An Algebraic Treatment*,
Studies in Fuzziness and Soft Computing 338, DOI 10.1007/978-3-319-30406-9_11

defined his generalized concepts in a complete residuated lattice L and set a general research problem (Q):

Do there exist a fuzzy set of logical axioms and a set of fuzzy rules of inference such that for any fuzzy theory $\mathcal{T}$ and any formula α the degree to which α follows from $\mathcal{T}$ equals exactly the degree to which α is provable from $\mathcal{T}$?

The answer depends on the set L of truth values; if it is affirmative then the corresponding logic enjoys *Pavelka-style completeness*. Pavelka himself limited to address the issue in the case L is a finite chain or the unit real interval $[0, 1]$ and proved (essentially) that this question has an affirmative answer if, and only if L is equipped with Łukasiewicz operations, i.e. an MV-algebra. In this sense Pavelka's logic—assuming L is the standard MV-algebra—is an extension of Łukasiewicz logic.

Our intention is to examine the issue when L is a perfect MV-algebra, in particular Chang's algebra. Since perfect MV-algebras are not complete we can have only partial generalizations and results. In the following brief review of the main concepts of Pavelka's general logic framework we follow mainly [7].

11.2 The Language $\mathcal{F}$ of Perfect Pavelka Logic

We start by assuming that the set L of truth values is the Chang algebra C and consider a zero order language with an infinitely countable set of propositional variables $\mathtt{p}, \mathtt{q}, \mathtt{r}, \ldots$, and two *truth constants* $\mathbf{0}, \mathbf{t}$. Propositional variables and the truth constant constitute the set $\mathcal{F}_0$ of *atomic formulas*. The elementary logical connectives are *implication* '`imp`' and *bold conjunction* '`and`'. The set $\mathcal{F}$ of all well formed formulas (wffs) is obtained in the natural way: atomic formulas are wffs and if α, β are wffs, then '$\alpha \ \mathtt{imp}\ \beta$', '$\alpha \ \mathtt{and}\ \beta$' are wffs. Other logical connectives are introduced as abbreviations; *negation* '`not`' is defined by setting $\mathtt{not}\,\alpha := \alpha \ \mathtt{imp}\ \mathbf{0}$, where $\mathbf{0}$ is the truth constant representing (absolute) falsity, and *bold disjunction* '`or`'. is defined by setting

$$\alpha \ \mathtt{or}\ \beta := (\mathtt{not}\,\alpha)\ \mathtt{imp}\ \beta.$$

Another connective $\underline{\mathtt{or}}$ called *weak disjunction* is an abbreviation $\alpha \ \underline{\mathtt{or}}\ \beta := (\alpha \ \mathtt{imp}\ \beta)\ \mathtt{imp}\ \beta$, and as usual, an *equivalence* is an abbreviation $\alpha \ \mathtt{equiv}\ \beta := (\alpha \ \mathtt{imp}\ \beta)\ \mathtt{and}\ (\beta \ \mathtt{imp}\ \alpha)$. We will also introduce the following abbreviations; for reasons that will reveal in the next chapter, also they will be called *truth constants*.

$\mathbf{1} := \mathtt{not}\,\mathbf{0}$, $\mathbf{f} := \mathtt{not}\,\mathbf{t}$,
$\mathbf{t}^2 := \mathbf{t}\ \mathtt{and}\ \mathbf{t}$, $\mathbf{f}^2 := \mathbf{f}\ \mathtt{or}\ \mathbf{f}$,

$\vdots$

$\mathbf{t}^n := \mathbf{t}^{n-1}\ \mathtt{and}\ \mathbf{t}$, $\mathbf{f}^n := \mathbf{f}^{n-1}\ \mathtt{or}\ \mathbf{f}$ for all $n > 2$.

The truth constant $\mathbf{1}$ corresponds to (absolute) truth, while the truth constant $\mathbf{t}$ has an intuitive meaning *quasi true*. Similarly the truth constant $\mathbf{f}$ has an intuitive meaning *quasi false*. Truth constants will be denoted by **a**, **b**, **c**. Usually logical connectives are denoted by the same symbols than their algebraic counterparts. For example $\rightarrow$ stands both for logical connective 'implication' and for algebraic residuation. This may cause something confusion and therefore our notation is close to the intuitive meaning of the logical connectives. However, to distinguish them from natural language words we write them by tt-fonts.

11.3 Semantics: Valuations

Semantics in Perfect Pavelka's Logic is introduced in the following way: any mapping $v : \mathcal{F}_0 \rightarrow C$ such that $v(\mathbf{0}) = 0$, $v(\mathbf{t}) = t \in C$ can be extended recursively into the whole $\mathcal{F}$ by setting

$$v(\alpha \,\texttt{imp}\, \beta) = v(\alpha) \rightarrow v(\beta) \quad \text{and} \quad v(\alpha \,\texttt{and}\, \beta) = v(\alpha) \odot v(\beta).$$

Such mappings v are called *valuations*. It is easy to see that for all valuations hold $v(\alpha \,\texttt{or}\, \beta) = v(\alpha) \oplus v(\beta)$, $v(\texttt{not}\, \alpha) = v(\alpha)^*$, $v(\mathbf{1}) = 1$, and for all natural n, $v(\mathbf{t}^n) = t^n$, $v(\mathbf{f}^n) = f^n$. Obviously any valuation v is a bijective mapping between the set of all truth constants **a** and elements $a \in C$; $v(\mathbf{a}) = a$.

In Pavelka's general setting, the *truth degree* of a wff α is the infimum of all values $v(\alpha)$, that is

$$\mathcal{C}^{\text{sem}}(\alpha) = \bigwedge \{v(\alpha) \mid v \text{ is a valuation}\},$$

whenever such an infimum exists in the truth value set L. However, in C this is not always the case as C is not complete as a lattice. Anyhow, if $\mathcal{C}^{\text{sem}}(\alpha)$ exists and is equal to a, we denote $\models_a \alpha$. In classical logic, (the axioms of) a *theory* is composed of a set of wffs assumed to be true. In order to define a *fuzzy theory*, we take $\mathcal{T} \subseteq \mathcal{F}$ and associate to each $\alpha \in \mathcal{T}$ a value $\mathcal{T}(\alpha)$ determining its degree of truth. We consider valuations v such that $\mathcal{T}(\alpha) \leq v(\alpha)$ for all wffs $\alpha \in \mathcal{T}$. If such a valuation exists, then $\mathcal{T}$ is called *satisfiable* and v *satisfies* $\mathcal{T}$. We say that the corresponding formulas α are the *special axioms* of the fuzzy theory $\mathcal{T}$ (called *non-logical axioms* of $\mathcal{T}$ in [4, 7]). Then we consider values

$$\mathcal{C}^{\text{sem}}(\mathcal{T})(\alpha) = \bigwedge \{v(\alpha) \mid v \text{ is a valuation, } v \text{ satisfies } \mathcal{T}\},$$

assuming such an infimum exists in C; if it exists and equals to a, we denote $\mathcal{T} \models_a \alpha$. Due to the linearity and discrete structure of C we observe that if the value $\mathcal{C}^{\text{sem}}(\mathcal{T})(\alpha)$ exists, then

$$\mathcal{C}^{\text{sem}}(\mathcal{T})(\alpha) = \min\{v(\alpha) \mid v \text{ is a valuation, } v \text{ satisfies } \mathcal{T}\} \in C.$$

Thus, if $\mathcal{C}^{\text{sem}}(\mathcal{T})(\alpha) = a$, then there is a valuation v satisfying $\mathcal{T}$ with $v(\alpha) = a$.

11.4 Syntax: Axioms and Rules of Inference

The logical axioms in Perfect Pavelka's Logic, denoted by A, are composed of the following twelve forms of formulas (axiomatic schemata):

(Ax1) $\alpha \texttt{ imp } \alpha$,
(Ax2) $(\alpha \texttt{ imp } \beta) \texttt{ imp } [(\beta \texttt{ imp } \gamma) \texttt{ imp } (\alpha \texttt{ imp } \gamma)]$,
(Ax3) $(\alpha_1 \texttt{ imp } \beta_1) \texttt{ imp } \{(\beta_2 \texttt{ imp } \alpha_2) \texttt{ imp } [(\beta_1 \texttt{ imp } \beta_2) \texttt{ imp } (\alpha_1 \texttt{ imp } \alpha_2)]\}$,
(Ax4) $\alpha \texttt{ imp } \mathbf{1}$,
(Ax5) $\mathbf{0} \texttt{ imp }$,
(Ax6) $(\alpha \texttt{ and not } \alpha) \texttt{ imp } \beta$,
(Ax7) $\mathbf{a} \texttt{ imp } \mathbf{b}$,
(Ax8) $\alpha \texttt{ imp } (\beta \texttt{ imp } \alpha)$,
(Ax9) $(\mathbf{1} \texttt{ imp } \alpha) \texttt{ imp } \alpha$,
(Ax10) $[(\alpha \texttt{ imp } \beta) \texttt{ imp } \beta] \texttt{ imp } [(\beta \texttt{ imp } \alpha) \texttt{ imp } \alpha]$,
(Ax11) $(\texttt{not } \alpha \texttt{ imp not } \beta) \texttt{ imp } (\beta \texttt{ imp } \alpha)$,
(Ax12) $[(\alpha \texttt{ or } \alpha) \texttt{ and } (\alpha \texttt{ or } \alpha)] \texttt{ equiv } [(\alpha \texttt{ and } \alpha) \texttt{ or } (\alpha \texttt{ and } \alpha)]$.

It is easy to very that all the axiomatic schemata δ in (Ax1)–(Ax6) and (Ax8)–(Ax12) are 1-tautologies, that is $\mathcal{C}^{\text{sem}}(\delta) = 1$ and, for axioms (Ax7), $\mathcal{C}^{\text{sem}}(\mathbf{a} \texttt{ imp } \mathbf{b}) = a \to b \in C$. A *many-valued rule of inference* is a schema

$$\frac{\alpha_1, \ldots, \alpha_n}{r^{\text{syn}}(\alpha_1, \ldots, \alpha_n)}, \quad \frac{a_1, \ldots, a_n}{r^{\text{sem}}(a_1, \ldots, a_n)}$$

where the wffs $\alpha_1, \ldots, \alpha_n$ are *premises* and the wff $r^{\text{syn}}(\alpha_1, \ldots, \alpha_n)$ is the *conclusion*. The values $a_1, \ldots, a_n, r^{\text{sem}}(a_1, \ldots, a_n) \in C$ are the corresponding *degrees*; for this reason, Pavelka's approach is sometimes called a *logic with evaluated syntax*. In two valued logic the degrees would all be equal to 1 corresponding to true premisses. The mappings $r^{\text{sem}} : C^n \curvearrowright C$ are assumed to satisfy isotonicity condition: if $a_k \leq b_k$, then

$$r^{\text{sem}}(a_1, \ldots, a_k, \ldots, a_n) \leq r^{\text{sem}}(a_1, \ldots, b_k, \ldots, a_n) \tag{11.1}$$

for each index $1 \leq k \leq n$. Moreover, many-valued rules are required to be *sound* in the sense that

$$r^{\text{sem}}(v(\alpha_1), \ldots, v(\alpha_n)) \leq v(r^{\text{syn}}(\alpha_1, \ldots, \alpha_n))$$

holds for all valuations v. The following are examples of many-valued rules of inference in any residuated lattice valued Pavelka style logic.

Generalized Modus Ponens (GMP):

$$\frac{\alpha, \alpha \texttt{ imp } \beta}{\beta}, \quad \frac{a, b}{a \odot b}$$

a-Consistency testing rules (**a**-CTR):

$$\frac{\mathbf{a}}{\mathbf{0}}\ ,\ \frac{b}{c}$$

where **a** is a truth constant and $c = 0$ if $b \leq a$ and $c = 1$ otherwise.

a-Lifting rules (**a**-LR):

$$\frac{\alpha}{\mathbf{a}\ \mathtt{imp}\ \alpha}\ ,\ \frac{b}{a \to b}$$

where **a** is a truth constant.

Rule of Bold Conjunction (RBC):

$$\frac{\alpha, \beta}{\alpha\ \mathtt{and}\ \beta}\ ,\ \frac{a, b}{a \odot b}$$

It is easy to see that also a Rule of Bold Disjunction (RBD, not included in the original list of Pavelka)

$$\frac{\alpha, \beta}{\alpha\ \mathtt{or}\ \beta}\ ,\ \frac{a, b}{a \oplus b}$$

is a rule of inference in Pavelka's sense in C-valued logic. Indeed, isotonicity of r^{sem} follows by the isotonicity of the MV-operation $\oplus$ and soundness can be verified by taking a valuation v and observing that

$$\begin{aligned} r^{\text{sem}}(v(\alpha), v(\beta)) &= v(\alpha) \oplus v(\beta) \\ &= v(\alpha\ \mathtt{or}\ \beta) \\ &= v(r^{\text{syn}}(\alpha, \beta)). \end{aligned}$$

These rules constitute a set $\mathtt{R}$. An $\mathtt{R}$-*proof* w of a wff α in a fuzzy theory $\mathcal{T}$ is a finite sequence

$$\begin{matrix} \alpha_1 & , & a_1 \\ \vdots & & \vdots \\ \alpha_m & , & a_m \end{matrix}$$

where

(i) $\alpha_m = \alpha$,

(ii) for each i, $1 \leq i \leq m$, α_i is a logical axiom or a special axiom of a fuzzy theory $\mathcal{T}$, or there is a many-valued rule of inference in $\mathtt{R}$ and well formed formulas $\alpha_{i_1}, \ldots, \alpha_{i_n}$ with $i_1, \ldots, i_n < i$ such that $\alpha_i = r^{\text{syn}}(\alpha_{i_1}, \ldots, \alpha_{i_n})$,

(iii) for each i, $1 \le i \le m$, the value $a_i \in C$ is given by $a_i =$

$$\begin{cases} a \to b & \text{if } \alpha_i \text{ is the axiom (Ax7) } \mathbf{a} \texttt{ imp } \mathbf{b}, \\ 1 & \text{if } \alpha_i \text{ is some other logical axiom in } \mathtt{A}, \\ \mathcal{T}(\alpha_i) & \text{if } \alpha_i \text{ is a special axiom of a fuzzy theory } \mathcal{T}, \\ r^{\text{sem}}(a_{i_1}, \ldots, a_{i_n}) & \text{if } \alpha_i = r^{\text{syn}}(\alpha_{i_1}, \ldots, \alpha_{i_n}). \end{cases}$$

The value a_m is called the *degree* of the R-proof w. Since a wff α may have various R-proofs with different degrees, we define the *provability degree* of a formula α to be the supremum of all such values, i.e.,

$$\mathcal{C}^{\text{syn}}(\mathcal{T})(\alpha) = \bigvee \{a_m \mid w \text{ is an R-proof for } \alpha \text{ in the fuzzy theory } \mathcal{T}\},$$

whenever such a supremum exists: we denote $\mathcal{T} \vdash_a \alpha$ if $\mathcal{C}^{\text{syn}}(\mathcal{T})(\alpha) = a$. Notice that such a value may not exist in C. In particular, $\mathcal{C}^{\text{syn}}(\mathcal{T})(\alpha) = 0$ means that the degree of any R-proof w of α is 0.

Again, due to the linearity and discrete structure of C we observe that if the value $\mathcal{C}^{\text{syn}}(\mathcal{T})(\alpha)$ exists, then

$$\mathcal{C}^{\text{syn}}(\mathcal{T})(\alpha) = \max\{a_m \mid a_m \text{ is the value of the R-proof } w \text{ for } \alpha \text{ in } \mathcal{T}\} \in C.$$

Consequently, if $\mathcal{C}^{\text{syn}}(\mathcal{T})(\alpha)$ exists and is equal to a, then there is an R-proof w for α with the value a.

Recall that Axiom Schemas (Ax8), (Ax2), (Ax10), (Ax11) are the axioms of Łukasiewicz propositional logic whose only rule of inference is Modus Ponens. Thus, all the formulas that are provable (it is the usual sense of the word, i.e. there is a classical proof for them) in Łukasiewicz propositional logic are provable at the highest degree also in the present logic.

Remark 11.1 **a**-Lifting Rules can be seen as particular instances of RBD. Indeed, since `not`**a** stands for (**a** `imp` **0**) and `not`**a** `or` α is an abbreviation for `not not`**a** `imp` α, we have the following R-proof for (**a** `imp` α)

`not`**a**	, a^*	,	(Ax7)
α	, b	,	Assumption
`not`**a** `or` α	, $a^* \oplus b$	,	RBD
`not not`**a** `imp` α	, $a^* \oplus b$	,	Abbreviation
(`not not`**a** `imp` α) `imp` (**a** `imp` α)	, 1	,	Łukasiewicz logic
(**a** `imp` α)	, $a \to b$	,	GMP, $a \to b = a^* \oplus b$

On the basis of the choice of the axioms and by soundness condition of rules of inference, a satisfiable fuzzy theory $\mathcal{T}$ is *sound*; if $\mathcal{T} \vdash_a \alpha$ and $\mathcal{T} \models_b \alpha$ hold, then $a \le b$.

We observe that any truth constant **a** has the following R-proof

$$\begin{array}{lll} (\mathbf{1}\ \texttt{imp}\ \mathbf{a})\ \texttt{imp}\ \mathbf{a} & , 1 & , (\text{Ax9}) \\ (\mathbf{1}\ \texttt{imp}\ \mathbf{a}) & , 1 \to a & , (\text{Ax7}) \\ \mathbf{a} & , a & , \text{GMP} \end{array}$$

as $1 \odot (1 \to a) = a$ in MV-algebras. In fact, in Pavelka's logic any formula α is provable, at least to a degree 0, this is denoted by $\mathcal{T} \vdash_0 \alpha$. Indeed, any well formed formula α has the following R-proof

$$\begin{array}{lll} \mathbf{0}\ \texttt{imp}\ [(\alpha\ \texttt{imp}\ \mathbf{0})\ \texttt{imp}\ \mathbf{0}] & , 1 & , (\text{Ax8}) \\ \mathbf{0} & , 0 & , (\text{Ax7}), (\text{Ax9}), \text{GMP} \\ (\alpha\ \texttt{imp}\ \mathbf{0})\ \texttt{imp}\ \mathbf{0} & , 0 & , \text{GMP} \\ [(\alpha\ \texttt{imp}\ \mathbf{0})\ \texttt{imp}\ \mathbf{0}]\ \texttt{imp}\ \alpha & , 1 & , \text{Łukasiewicz logic} \\ \alpha & , 0 & , \text{GMP} \end{array}$$

This leads us to the following

Definition 11.2 A fuzzy theory $\mathcal{T}$ is *consistent* if $\mathcal{C}^{syn}(\mathcal{T})(\mathbf{a}) = a$ for all truth constants **a**, otherwise $\mathcal{T}$ is *inconsistent*.

Then we have

Proposition 11.3 *A fuzzy theory $\mathcal{T}$ is inconsistent iff $\mathcal{T} \vdash_1 \alpha$ holds for any wff α.*

Proof Assume $\mathcal{T}$ is inconsistent. Then there is a truth constant **a** and an R-proof with value b such that $a < b = \mathcal{C}^{syn}(\mathcal{T})(\mathbf{a})$. Then for any wff α we have

$$\begin{array}{lll} \mathbf{a} & , b & , \text{Assumption} \\ \mathbf{0} & , 1 & , \mathbf{a}\text{-CTR} \\ \mathbf{0}\ \texttt{imp}\ \alpha & , 1 & , (\text{Ax5}) \\ \alpha & , 1 & , \text{GMP} \end{array}$$

We conclude that $\mathcal{T} \vdash_1 \alpha$ holds. Conversely, if $\mathcal{T} \vdash_1 \alpha$ holds for any wff α, then in particular $\mathcal{T} \vdash_1 \mathbf{0}$ and $0 \neq 1$. □

Proposition 11.4 *A fuzzy theory $\mathcal{T}$ is inconsistent iff the following condition holds:*

(*C*) *There is a wff α and* R*-proofs w, w' with values a, b for α and* `not` *α, respectively, such that $0 < a \odot b$.*

Proof Let (C) hold and β be an arbitrary wff. Then there is the following R-proof for β in $\mathcal{T}$.

α	, a	, Assumption
$\texttt{not}\,\alpha$	, b	, Assumption
$\texttt{not}\,\alpha\,\texttt{and}\,\alpha$	, $a \odot b$	, RBC
$(\texttt{not}\,\alpha\,\texttt{and}\,\alpha)\ \texttt{imp}\ \mathbf{0}$	, 1	, (Ax6)
$\mathbf{0}$	, $a \odot b$	, GMP
$\mathbf{0}$	, 1	, **0**-CTR
$\mathbf{0}\ \texttt{imp}\ \beta$	, 1	, (Ax5)
β	, 1	, GMP

If conversely $\mathcal{T}$ is inconsistent, then $\mathcal{T} \vdash_1 \alpha$ and $\mathcal{T} \vdash_1 \texttt{not}\,\alpha$ for any wff α. Thus there are R-proofs w, w' with values 1, 1 for α and $\texttt{not}\,\alpha$, respectively, and trivially $0 < 1 \odot 1$. □

Proposition 11.5 *A satisfiable fuzzy theory $\mathcal{T}$ is consistent.*

Proof Let v satisfy $\mathcal{T}$ and $v(\alpha) = c$, where α is a wff. Then $v(\texttt{not}\,\alpha) = c^*$. If w, w' are R-proofs with values a, b for α and $\texttt{not}\,\alpha$, respectively, then by soundness $a \leq c$ and $b \leq c^*$. Therefore $a \odot b \leq c \odot c^* = 0$. Thus, $\mathcal{T}$ is not inconsistent and is therefore consistent. □

Proposition 11.6 *If $\mathcal{T} \vdash_a \alpha$ then $\mathcal{T} \vdash_1 (\mathbf{a}\ \texttt{imp}\ \alpha)$.*

Proof If $\mathcal{T} \vdash_a \alpha$ then there is the following R-proof in $\mathcal{T}$:

α	, a	, Assumption
$\mathbf{a}\ \texttt{imp}\ \alpha$	, 1	, **a**-LR

Therefore $\mathcal{T} \vdash_1 (\mathbf{a}\ \texttt{imp}\ \alpha)$. □

11.5 $\mathcal{T}$-complete Formulas in Perfect Pavelka Logic

Since perfect MV-algebras are not complete when considered as lattices, Pavelka's ideas cannot be applied as such in perfect MV-algebra framework. However, given a fuzzy theory $\mathcal{T}$ which might not be complete in Pavelka's sense, there are still interesting formulas α that satisfy

$$\mathcal{C}^{syn}(\mathcal{T})(\alpha) = \mathcal{C}^{sem}(\mathcal{T})(\alpha); \tag{11.2}$$

call them $\mathcal{T}$*-complete formulas*. We limit to *strong fuzzy theories* $\mathcal{T}$; the special axioms are given in the form

$$\mathcal{T}(\alpha) = a\,\text{and}\,\mathcal{T}(\alpha\ \texttt{imp}\ \mathbf{a}) = 1,$$

where α is a well-formed formula, $\mathbf{a}$ is a truth constant and a is the corresponding value in the truth value set L. We give an affirmative answer to Pavelka's a general question (Q) with respect to strong satisfiable fuzzy theories and certain subsets of formulas. The only new algebraic result needed is the following and presented in [8].

Remark 11.7 The unique solution of $a \to x = b \neq \mathbf{1}$, where a, b are elements of an MV-algebra, is $x = a \odot b$.

Before considering Pavelka's question (Q), we show that Lindenbaum-Tarski algebra is available in Perfect Pavelka framework; we define on the set $\mathcal{F}$ of well-formed formula a binary relation $\preceq$ by setting

$$\alpha \preceq \beta \text{ if, and only if } \mathcal{T} \vdash_1 (\alpha \; \texttt{imp} \; \beta).$$

By (Ax1), (Ax2) and GMP it is easy to show that $\preceq$ is reflexive and transitive, and therefore, a quasi-order. Hence, by defining a binary relation $\equiv$ via

$$\alpha \equiv \beta \text{ if, and only if } \mathcal{T} \vdash_1 (\alpha \; \texttt{imp} \; \beta) \text{ and } \mathcal{T} \vdash_1 (\beta \; \texttt{imp} \; \alpha).$$

we obtain an equivalence relation on $\mathcal{F}$. As usual, we denote by $|\alpha|$ the equivalence class defined by α and the set of all equivalence classes by $\mathcal{F}/\equiv$. The relation $\equiv$ is a congruence with respect to the logical connective imp. Indeed, assume $\alpha_1 \equiv \beta_1$ and $\alpha_2 \equiv \beta_2$. Then $\mathcal{T} \vdash_1 \alpha_1 \; \texttt{imp} \; \beta_1$ and $\mathcal{T} \vdash_1 \beta_2 \; \texttt{imp} \; \alpha_2$. Let

$$\gamma = [(\beta_1 \; \texttt{imp} \; \beta_2) \; \texttt{imp} \; (\alpha_1 \; \texttt{imp} \; \alpha_2)].$$

Then we have the following R-proof

$$\begin{array}{ll}
(\alpha_1 \; \texttt{imp} \; \beta_1) & , 1 \text{ , Assumption} \\
(\alpha_1 \; \texttt{imp} \; \beta_1) \; \texttt{imp} \; [(\beta_2 \; \texttt{imp} \; \alpha_2) \; \texttt{imp} \; \gamma] & , 1 \text{ , (Ax3)} \\
(\beta_2 \; \texttt{imp} \; \alpha_2) \; \texttt{imp} \; \gamma & , 1 \text{ , GMP} \\
(\beta_2 \; \texttt{imp} \; \alpha_2) & , 1 \text{ , Assumption} \\
\gamma & , 1 \text{ , GMP}
\end{array}$$

We conclude $\mathcal{T} \vdash_1 (\beta_1 \; \texttt{imp} \; \beta_2) \; \texttt{imp} \; (\alpha_1 \; \texttt{imp} \; \alpha_2)$. In a similar manner we prove that $\mathcal{T} \vdash_1 (\alpha_1 \; \texttt{imp} \; \alpha_2) \; \texttt{imp} \; (\beta_1 \; \texttt{imp} \; \beta_2)$. Therefore

$$(\alpha_1 \; \texttt{imp} \; \alpha_2) \equiv (\beta_1 \; \texttt{imp} \; \beta_2).$$

In particular, if $\alpha \equiv \beta$, then $\texttt{not} \; \alpha \equiv \texttt{not} \; \beta$, since $\mathbf{0} \equiv \mathbf{0}$ and $\texttt{not} \; \alpha = \alpha \; \texttt{imp} \; \mathbf{0}$ by definition. Accordingly, the equations

$$|\alpha| \to |\beta| = |\alpha \; \texttt{imp} \; \beta| \text{ and } |\alpha|^* = |\, \texttt{not} \; \alpha|$$

define a binary and unary operations, respectively, in $\mathcal{F}/\equiv$. By (Ax5) and (Ax4), for any equivalence class $|\alpha|$ holds $|\mathbf{0}| \leq |\alpha| \leq |\mathbf{1}|$, where the order in $\mathcal{F}/\equiv$ given by

$$|\alpha| \leq |\beta| \text{ if, and only if } \mathcal{T} \vdash_1 (\alpha \; \texttt{imp} \; \beta).$$

We also observe

Proposition 11.8 *$\mathcal{T} \vdash_1 \alpha$ if, and only if $|\alpha| = |\mathbf{1}|$.*

Proof Indeed, if $|\alpha| = |\mathbf{1}|$ then $\mathcal{T} \vdash_1 \mathbf{1}\,\texttt{imp}\,\alpha$; use (Ax9) and GMP to obtain $\mathcal{T} \vdash_1 \alpha$. Conversely, if $\mathcal{T} \vdash_1 \alpha$ then, by Proposition 11.6, $\mathcal{T} \vdash_1 \mathbf{1}\ \texttt{imp}\ \alpha$, hence $|\mathbf{1}| \leq |\alpha|$ whence $|\mathbf{1}| = |\alpha|$. □

Proposition 11.9 *For a consistent fuzzy theory $\mathcal{T}$, the Lindenbaum-Tarski algebra $\langle \mathcal{F}/\equiv, \rightarrow, {}^*, |\mathbf{1}|\rangle$ is a Wajsberg algebra and, hence an MV-algebra. In fact, the Lindenbaum-Tarski algebra is a perfect MV-algebra.*

Proof Let α, β, γ be wffs. We observe that Wajsberg axioms hold in the algebra $\langle \mathcal{F}/\equiv, \rightarrow, {}^*, |\mathbf{1}|\rangle$:
1° by (Ax8) and (Ax9), $|\alpha| = |\mathbf{1}| \rightarrow |\alpha|$,
2° by (Ax2), $(|\alpha| \rightarrow |\beta|) \rightarrow [(|\beta| \rightarrow |\gamma|) \rightarrow (|\alpha| \rightarrow |\gamma|)] = |\mathbf{1}|$,
3° by (Ax10), $(|\alpha| \rightarrow |\beta|) \rightarrow |\beta| = (|\beta| \rightarrow |\alpha|) \rightarrow |\alpha|$,
4° by (Ax11), $(|\alpha|^* \rightarrow |\beta|^*) \rightarrow (|\beta| \rightarrow |\alpha|) = |\mathbf{1}|$.

Therefore $\langle \mathcal{F}/\equiv, \rightarrow, {}^*, |\mathbf{1}|\rangle$ can be seen as an MV-algebra; use (Ax12) to show that Eq. 4.1 is satisfied in $\mathcal{F}/\equiv$. □

Now we return to Pavelka's question (Q); we have the following partial solution.[1]

Theorem 11.10 (A Set of $\mathcal{T}$-complete Formulas) *Assume a strong fuzzy theory $\mathcal{T}$ is satisfiable. Assume in all R-proofs of a formula α, in all instances where GMP is used holds either $a = b = 1$ or $b \neq 1$. Then α is $\mathcal{T}$-complete.*

Proof For any valuation v that satisfies $\mathcal{T}$ holds $a = \mathcal{T}(\alpha) \leq v(\alpha)$ and

$$1 = \mathcal{T}(\alpha\ \texttt{imp}\ \mathbf{a}) \leq v(\alpha\ \texttt{imp}\ \mathbf{a}) = v(\alpha) \rightarrow a \leq 1.$$

Therefore $v(\alpha) \leq v(\mathbf{a}) = a$ and so $v(\alpha) = a$. All the special axioms $\mathcal{T}(\alpha)$ and $\mathcal{T}(\alpha\ \texttt{imp}\ \mathbf{a})$ have a trivial R-proof at the corresponding degree a and 1, respectively. Similarly, all the logical axioms (Ax1)–(Ax6), and (Ax8)–(Ax12) have a trivial proof at the degree 1 and they are 1-tautologies. Finally, by axioms (Ax7); for all truth constants $\mathbf{a}, \mathbf{b}$ a formula ($\mathbf{a}\ \texttt{imp}\ \mathbf{b}$) is an axiom of degree $a \rightarrow b$ and satisfies $v(\mathbf{a}\ \texttt{imp}\ \mathbf{b}) = a \rightarrow b$. By soundness, the completeness condition

$$C^{sem}(\mathcal{T})(\alpha) = C^{syn}(\mathcal{T})(\alpha) \tag{11.3}$$

holds for all logical and special axioms α.

Now it it easy to prove inductively that the completeness condition holds for all well formed formulas α assuming that in all instances where GMP is used holds either $a = b = 1$ or $b \neq 1$. Indeed, assume α and β have R-proof of degree a and

[1] In [9] there is another approach to Perfect Pavelka Logic. However, it seems that there is a gap in the proof of Proposition 16.

b, respectively and there is a valuation v such that $v(\alpha) = a$ and $v(\beta) = b$, then using the rules RBC, RBD and a-CTR, respectively, α and β, α or β and $\mathbf{0}$ have a R-proof of degree $a \odot b$, $a \oplus b$ and 0, respectively, and $v(\alpha \texttt{ and } \beta) = a \odot b$, $v(\alpha \texttt{ or } \beta) = a \oplus b$ and $v(\mathbf{0}) = 0$. Finally, assume α and α imp β have R-proof of degree a and $b \neq 1$, respectively and there is a valuation v such that $v(\alpha) = a$ and $v(\alpha \texttt{ imp } \beta) = b$, then using the rule GMP, β has a R-proof with a degree $a \odot b$ and $v(\beta) = a \odot b$. Obviously the claim holds if $a = b = 1$. The proof is complete. □

Notice that the Theorem above gives sufficient, and not necessary conditions; there are satisfiable fuzzy theories $\mathcal{T}$ that are not strong and still (11.3) holds for some formulas α.

11.6 Examples and New Rules of Inference

Consider the following realistic but invented

Problem *Prolonged and constant hurry often leads to nervousness, and often poor eating habits cause peritonitis. Gastric ulcer, in turn, is always caused by nervousness or peritonitis. The severity of the disease increases with age; for an elderly person gastric ulcer can be a fatal disease. Mr. A is constantly in a hurry and he eats unhealthy. Will he contract gastric ulcer? If yes, will the disease be fatal?*

Solution Let us consider the situation in Perfect Pavelka logic framework and introduce the following entries

p stands for *A is always in a hurry*,
q stands for *A is nervous*,
r stands for *A has poor eating habits*,
s stands for *A contracts peritonitis*,
t stands for *A contracts gastric ulcer*,
w stands for *A is old*,
z stands for *The illness is fatal*.

Then construct a fuzzy theory $\mathcal{T}$ whose special axioms are

$\mathcal{T}(\texttt{p}) = 1$,
$\mathcal{T}(\texttt{r}) = 1$,
$\mathcal{T}(\texttt{p imp q}) = t^3$ (since the implication is *often*, but not always),
$\mathcal{T}(\texttt{r imp s}) = t^3$ (since the implication is *often*, but not always),
$\mathcal{T}((\texttt{q or s}) \texttt{ imp t}) = 1$,
$\mathcal{T}(\texttt{w imp (t imp z)}) = 1$,
$\mathcal{T}(\texttt{w}) = t^5$ (the degree by which Mr. A belongs to the fuzzy set Old person).

The first task is to check whether the fuzzy theory $\mathcal{T}$ meaningful, i.e. whether it is satisfiable. After searching for a while, we find the following valuation v:

$$v(\texttt{p}) = v(\texttt{r}) = v(\texttt{t}) = 1, v(\texttt{q}) = v(\texttt{s}) = t^3, v(\texttt{w}) = v(\texttt{z}) = t^5.$$

It is easy to see that this valuation satisfies $\mathcal{T}$. Then we look for R-proofs for t, consider the following

```
p                , 1  , Special axiom
p imp q          , t^3 , Special axiom
q                , t^3 , GMP
r                , 1  , Special axiom
r imp s          , t^3 , Special axiom
s                , t^3 , GMP
q or s           , 1  , RBD
(q or s) imp t   , 1  , Special axiom
t                , 1  , GMP
```

Thus, the conclusion *A contracts gastric ulcer* is absolutely true. It remains to clarify how fatal the disease is. Since we already found a valuation v that satisfies $\mathcal{T}$ and $v(\mathtt{z}) = t^5$, the sentence *The illness is fatal* can be true and provable maximally at a degree t^5. We have the following R-proof for z

```
w                    , t^5 , Special axiom
w imp (t imp z)      , 1   , Special axiom
t imp z              , t^5 , GMP
t                    , 1   , Assumption
z                    , t^5 , GMP
```

We conclude that the sentence z is provable and valid at the degree t^5. Freely speaking, *Mr. A, a middle-aged man who is constantly in a hurry and has poor eating habits will contract gastric ulcer. However, the disease will not be completely fatal.*

In real life applications, finding an R-proof for some particular formula might be difficult and time-consuming. One solution is to try find first a classical proof and then extend it to a graded R-proof. This can be done without difficulties as Pavelka's general definition for a many-valued rule of inference and R-proof has a consequence that any classical logic rule of inference has a many-valued counterpart and each classical proof of a formula α has a graded proof. This of course does not mean that a formula that is provable in classical would be 1-provable in Perfect Pavelka Logic or even in the original [0, 1]-valued Pavelka logic. On the other hand, adding new rules that satisfy Pavelka's isotonic and soundness condition to a satisfiable and hence consistent system does not expand the set of provable sentences nor does it increase the provability degree of any well formed formula α. Next, we present a wide range of isotone and sound rules of inference. We extend the definition of R-proof correspondingly.

Proposition 11.11 *In any MV-algebra valued Pavelka style Fuzzy Logic the following schemas are many-valued rules of inference.*

Generalized Modus Tollendo Tollens (GMTT):

$$\frac{\texttt{not}\,\beta, (\alpha\ \texttt{imp}\ \beta)}{\texttt{not}\,\alpha}, \frac{a, b}{a \odot b}$$

Generalized Hypothetical Syllogism (GHS):

$$\frac{(\alpha\ \texttt{imp}\ \beta), (\beta\ \texttt{imp}\ \gamma)}{\alpha\ \texttt{imp}\ \gamma}, \frac{a, b}{a \odot b}$$

Generalized Commutative Law 1 GCL1:

$$\frac{\alpha\ \texttt{and}\ \beta}{\beta\ \texttt{and}\ \alpha}, \frac{a}{a}$$

Generalized Commutative Law 2 (GCL2):

$$\frac{\alpha\ \texttt{or}\ \beta}{\beta\ \texttt{or}\ \alpha}, \frac{a}{a}$$

Generalized Equivalence Law 1 (GEL1):

$$\frac{\alpha\ \texttt{equiv}\ \beta}{\alpha\ \texttt{imp}\ \beta}, \frac{a}{a}$$

Generalized Equivalence Law 2 (GEL2):

$$\frac{\alpha\ \texttt{equiv}\ \beta}{\beta\ \texttt{imp}\ \alpha}, \frac{a}{a}$$

Generalized Equivalence Law 3 (GEL3):

$$\frac{(\alpha\ \texttt{imp}\ \beta), (\beta\ \texttt{imp}\ \alpha)}{\alpha\ \texttt{equiv}\ \beta}, \frac{a, b}{a \wedge b}$$

Generalized Simplification Law 1 (GSL1):

$$\frac{\alpha\ \texttt{and}\ \beta}{\alpha}, \frac{a}{a}$$

Generalized Simplification Law 2 (GSL2):

$$\frac{\alpha\ \texttt{and}\ \beta}{\beta}, \frac{a}{a}$$

Generalized Rule of Introduction of Double Negation (GIDN):

$$\frac{\alpha}{\texttt{not}(\texttt{not}\,\alpha)}, \frac{a}{a}$$

Generalized Rule of Elimination of Double Negation (GEDN):

$$\frac{\texttt{not}(\texttt{not}\,\alpha)}{\alpha}, \frac{a}{a}$$

Generalized De Morgan Law 1 (GDML1):

$$\frac{(\texttt{not}\,\alpha)\,\texttt{and}\,(\texttt{not}\,\beta)}{\texttt{not}(\alpha\,\texttt{or}\,\beta)}, \frac{a}{a}$$

Generalized De Morgan Law 2 (GDML2):

$$\frac{\texttt{not}(\alpha\,\texttt{or}\,\beta)}{(\texttt{not}\,\alpha)\,\texttt{and}\,(\texttt{not}\,\beta)}, \frac{a}{a}$$

Generalized De Morgan Law 3 (GDML3):

$$\frac{(\texttt{not}\,\alpha)\,\texttt{or}\,(\texttt{not}\,\beta)}{\texttt{not}(\alpha\,\texttt{and}\,\beta)}, \frac{a}{a}$$

Generalized De Morgan Law 4 (GDML4):

$$\frac{\texttt{not}(\alpha\,\texttt{and}\,\beta)}{(\texttt{not}\,\alpha)\,\texttt{or}\,(\texttt{not}\,\beta)}, \frac{a}{a}$$

Generalized Addition Law (GAL):

$$\frac{\alpha}{\alpha\,\texttt{or}\,\beta}, \frac{a}{a}$$

Generalized Modus Tollendo Ponens (GMTP):

$$\frac{\texttt{not}\,\beta, (\alpha\,\texttt{or}\,\beta)}{\alpha}, \frac{a, b}{a \odot b}$$

Generalized Disjunctive Syllogism (GDS):

$$\frac{(\alpha\,\texttt{or}\,\beta), (\alpha\,\texttt{imp}\,\gamma), (\beta\,\texttt{imp}\,\delta)}{\gamma\,\texttt{or}\,\delta}, \frac{a, b, c}{a \odot b \odot c}$$

Generalized Rule of Introduction of Implication (GII):

$$\frac{\texttt{not}\,\alpha\,\texttt{or}\,\beta}{\alpha\,\texttt{imp}\,\beta}\,,\ \frac{a}{a}$$

Generalized Rule of Elimination of Implication (GEI):

$$\frac{\alpha\,\texttt{imp}\,\beta}{\texttt{not}\,\alpha\,\texttt{or}\,\beta}\,,\ \frac{a}{a}$$

Proof Isotonicity of the rules GMTT, GHS, GMTP, GDS and GEL3 follows from the fact that the operations $\odot$ and $\wedge$ are isotone. The other rules are trivially isotone. Soundness of most of these rules is a direct consequence from properties on valuation. We establish here soundness of GMTT and GMTP. GDS is left as an exercise for the reader. Assume v is a valuation. Then

$$\begin{aligned} r^{sem}(v(\texttt{not}\,\beta), v(\alpha\,\texttt{imp}\,\beta)) &= v(\texttt{not}\,\beta) \odot v(\alpha\,\texttt{imp}\,\beta)\\ &= v(\beta)^* \odot [v(\alpha) \to v(\beta)]\\ &= [v(\alpha) \to v(\beta)] \odot [v(\beta) \to \mathbf{0}]\\ &\leq v(\alpha) \to \mathbf{0}\\ &= v(\texttt{not}\,\alpha)\\ &= v(r^{syn}(\texttt{not}\,\beta, [\alpha\,\texttt{imp}\,\beta])). \end{aligned}$$

Thus, GMTT is sound. For GMTP we have

$$\begin{aligned} r^{sem}(v(\texttt{not}\,\beta), v(\alpha\,\texttt{or}\,\beta)) &= v(\texttt{not}\,\beta) \odot v(\alpha\,\texttt{or}\,\beta)\\ &= v(\beta)^* \odot [v(\alpha) \oplus v(\beta)]\\ &= v(\beta)^* \odot [v(\beta)^* \to v(\alpha)]\\ &\leq v(\alpha)\\ &= v(r^{syn}(\texttt{not}\,\beta, [\alpha\,\texttt{or}\,\beta])). \end{aligned}$$

□

All these rules are graded generalizations of classical rules of inference, as is already clear from their names. Here we give still three more.

New Rule 1:

$$\frac{\alpha\,\texttt{imp}\,\gamma, \alpha\,\underline{\texttt{or}}\,\beta}{\gamma\,\underline{\texttt{or}}\,\beta}\,,\ \frac{a, b}{a \odot b}$$

New Rule 2:

$$\frac{\mathbf{0}\,\underline{\texttt{or}}\,\alpha}{\alpha}\,,\ \frac{a}{a}$$

New Rule 3:

$$\frac{\alpha \,\texttt{imp}\, \gamma, \beta \,\texttt{imp}\, \gamma}{(\alpha \,\underline{\texttt{or}}\, \beta) \,\texttt{imp}\, \gamma}, \quad \frac{a, b}{a \odot b}$$

Recall that $\alpha \,\underline{\texttt{or}}\, \beta$ is an abbreviation for $(\alpha \,\texttt{imp}\, \beta) \,\texttt{imp}\, \beta$ and for any valuation v, $v(\alpha \,\underline{\texttt{or}}\, \beta) = v(\alpha) \vee v(\beta)$ holds.

The isotonicity of these rules follows by the isotonicity of $\odot$ in any residuated lattice. To prove that the New Rule 1 is sound, let v be a valuation. We reason

$$\begin{aligned}
r^{\text{sem}}(v(\alpha \,\texttt{imp}\, \gamma), v(\alpha \,\underline{\texttt{or}}\, \beta)) &= v(\alpha \,\texttt{imp}\, \gamma) \odot v(\alpha \,\underline{\texttt{or}}\, \beta)) \\
&= [v(\alpha) \to v(\gamma)] \odot [v(\alpha) \vee v(\beta)] \\
&= [(v(\alpha) \to v(\gamma)) \odot v(\alpha)] \vee [(v(\alpha) \to v(\gamma)) \odot v(\beta)] \\
&\leq v(\gamma) \vee v(\beta) \\
&= v(\gamma \,\underline{\texttt{or}}\, \beta) \\
&= v(r^{\text{syn}}(\alpha \,\texttt{imp}\, \gamma, \alpha \,\underline{\texttt{or}}\, \beta)).
\end{aligned}$$

Soundness of the New Rule 2 is obvious. To prove that the New Rule 3 is sound, let v be a valuation. Then

$$\begin{aligned}
r^{\text{sem}}(v(\alpha \,\texttt{imp}\, \gamma, \beta \,\texttt{imp}\, \gamma)) &= [v(\alpha) \to v(\gamma)] \odot [v(\beta) \to v(\gamma)] \\
&\leq [v(\alpha) \vee v(\beta)] \to v(\gamma) \\
&= v((\alpha \,\underline{\texttt{or}}\, \beta) \,\texttt{imp}\, \gamma) \\
&= v(r^{\text{syn}}(\alpha \,\texttt{imp}\, \gamma, \beta \,\texttt{imp}\, \gamma)).
\end{aligned}$$

Problem Next consider another example originally taken from [10] in classical logic context; the task is to study the validity of the following reasoning in classical logic, in Perfect Pavelka Logic and in the original Pavelka's [0, 1]-valued logic.

If there is no government subsidies of agriculture, then there are government controls of agriculture. If there are government controls of agriculture, then there is no agricultural depression. There is either an agricultural depression or overproduction. As a matter of fact, there is no overproduction. Therefore, there are government subsidies of agriculture.

Solution The special axioms of a corresponding crisp theory $\mathcal{T}$ are

$$\mathcal{T}(\texttt{not p imp q}) = 1, \mathcal{T}(\texttt{q imp not r}) = 1, \mathcal{T}(\texttt{r or s}) = 1, \text{ and } \mathcal{T}(\texttt{not s}) = 1,$$

where $\texttt{p}$, $\texttt{q}$, $\texttt{r}$, and $\texttt{s}$ abbreviate *There is government subsidies of agriculture*, *There are government controls of agriculture*, *There is agricultural depression*, and *There is an agricultural overproduction*, respectively. The formula $\texttt{p}$ is provable from the special axioms of $\mathcal{T}$ and classical logic (CL) as follows

(not p imp q) imp [(q imp not r) imp (not p imp not r)]	, 1 ,	Provable in CL
not p imp q	, 1 ,	Special axiom
(q imp not r) imp (not p imp not r)	, 1 ,	Modus Ponens
(q imp not r)	, 1 ,	Special axiom
not p imp not r	, 1 ,	Modus Ponens
(not p imp not r) imp (r imp p)	, 1 ,	Provable in CL
r imp p	, 1 ,	Modus Ponens
r or s	, 1 ,	Special axiom
not s	, 1 ,	Special axiom
r	, 1 ,	MTP
p	, 1 ,	Modus Ponens

where MTP stands for the classical inference rule Modus Tollendo Ponens. Now assume the special axioms are true, but only to a degree, say $\mathcal{T}(\texttt{not}\, p \;\texttt{imp}\; q) = t^3$, $\mathcal{T}(q \;\texttt{imp not}\; r) = t^2$, $\mathcal{T}(r \;\texttt{or}\; s) = t^4$ and $\mathcal{T}(\texttt{not}\, s) = t$. Thus we have a fuzzy theory in Perfect Pavelka Logic. The above classical proof can be transferred into an R-proof for p as follows

(not p imp q) imp [(q imp not r) imp (not p imp not r)]	, 1 ,	(Ax2)
not p imp q	, t^3 ,	Special axiom
(q imp not r) imp (not p imp not r)	, t^3 ,	GMP
(q imp not r)	, t^2 ,	Special axiom
not p imp not r	, t^5 ,	GMP
(not p imp not r) imp (r imp p)	, 1 ,	(Ax11)
r imp p	, t^5 ,	GMP
r or s	, t^4 ,	Special axiom
not s	, t ,	Special axiom
r	, t^5 ,	GMTP
p	, t^{10} ,	GMP

where we used Generalized Modus Tollendo Ponens GMTP. We conclude that p is provable at least to a degree t^{10}. Since for a valuation v such that $v(\texttt{p}) = t^{10}$, $v(\texttt{q}) = f^7$, $v(\texttt{r}) = t^5$, and $v(\texttt{s}) = f$ satisfies $v(\texttt{not p imp q}) = t^{10} \oplus f^7 = t^3$, $v(\texttt{q imp not r}) = t^7 \oplus f^5 = t^2$, $v(\texttt{r or s}) = t^5 \oplus f = t^4$, and $v(\texttt{not s}) = t$, we conclude that the fuzzy theory $\mathcal{T}$ is satisfiable and $\mathcal{C}^{\text{sem}}\mathcal{T}(\texttt{p}) = \mathcal{C}^{\text{syn}}\mathcal{T}(\texttt{p}) = t^{10}$.

We realize that from (at least partially) true premises the conclusion is also (at least partially) true in Perfect Pavelka Logic. This is not the case in the original [0, 1]-valued Pavelka Logic. Indeed, replace the special axioms by

$$\mathcal{T}(\texttt{not p imp q}) = 0.7, \mathcal{T}(\texttt{q imp not r}) = 0.8, \mathcal{T}(\texttt{r or s}) = 0.6, \text{ and } \mathcal{T}(\texttt{not s}) = 0.9.$$

Then a valuation v such that $v(\texttt{p}) = 0$, $v(\texttt{q}) = 0.7$, $v(\texttt{r}) = 0.5$, and $v(\texttt{s}) = 0.1$ satisfies $\mathcal{T}$ and $\mathcal{C}^{\text{sem}}\mathcal{T}(p) = \mathcal{C}^{\text{syn}}\mathcal{T}(\texttt{p}) = 0$.

11.7 What Can and What Cannot Be Expressed in Rational or Perfect Pavelka Logic

One of the arguments for (some) fuzzy logics is that they admit to explain the Sorites Paradox [11]. A fuzzy statement like 'this person is not bald' may be true at the beginning; then its truth degree may decrease by small decrements and after many repetitions, it may become false. The gradual true with a dense set of truth values admits to model this phenomenon and overcome a paradox from classical logic. Rational Pavelka logic, whose semantics is based on the standard MV-algebra, is one of the fuzzy logics where this can be explained.

In contrast to this, Chang's MV-algebra does not allow to explain the Sorites Paradox. If the statement 'this person is not bald' is true at the beginning and its truth degree decreases by small decrements (infinitesimals), it never reaches the value false; the truth degree remains infinitesimally close to 1 unless it makes a 'big jump' into the degrees infinitesimally close to 0.

Despite this difference, Chang's MV-algebra and infinitesmals have their role in modeling human reasoning. Example, let us consider the task of traveling to an airport. A truth degree should express how convenient the chosen way is. We may optimize the way in terms of cost, time, choosing the route, etc. However, all these improvements are negligible in comparison with the crucial question whether we catch our flight or not. Any number of 'small' advantages cannot compensate the big disadvantage when we miss the flight. Thus their representation by infinitesimals is adequate. This does not mean that—as soon as we choose only from options in which we catch the flight—we should ignore these small contributions, e.g., saving cost by choosing among several sufficiently fast options.

Thus the adequacy of both models depends on the specifics of the situation. Sometimes 'many small contributions may compensate a big change', sometimes not. More exactly, the use of infinitesimals admits to express that some changes (of a truth value) are nonzero, but infinitely many times smaller than others. The two semantics studied here are two extremes: The standard MV-algebra does not admit any infinitesimals, while the Chang's MV-algebra contains only infinitesimals (and their duals). As we have seen in this book, there are more general perfect MV-algebras which combine infinitesimals and non-infinitesimals. The semantics based on such MV-algebras could describe two types of changes of truth values—'big' ones which may model the Sorites Paradox and infinitesimal ones for which this paradox applies (as in classical logic). We expect that a Pavelka-style logic could be based on general perfect MV-algebras as well.

A typical example is the interval $[(0,0),(1,0)]$ in the lexicographical product $M = \mathbb{Q}\times_{\text{lex}}\mathbb{Z}$, where $\mathbb{Q}$, resp. $\mathbb{Z}$, is the set of all rational, resp. integer, numbers. The MV-algebraic operations on M are defined as follows:

$$\begin{aligned} \mathbf{0} &= (0, 0), \\ \mathbf{1} &= (1, 0), \\ (q, n)^* &= (1 - q, -n), \\ (p, k) \oplus (q, n) &= min((1, 0), (p + q, k + n)). \end{aligned}$$

The set $\{(0, n) | n \in \mathbb{Z}, n \geq 0\}$ is closed under $\oplus$, its elements are infinitesimals. On the other hand, elements of the form $(q, n), q > 0$, are not infinitesimals; the sum of $\lceil \frac{1}{q} \rceil$ such elements is $\mathbf{1} = (1, 0)$. The former elements have properties described in Perfect Pavelka Logic. The latter elements have properties known from Rational Pavelka Logic. We expect that the combination of both approaches could further extend the possibility of modeling of the human reasoning based on graduate truth values.

References

1. Hájek, P.: Metamathematics of Fuzzy Logic. Kluwer, Dordrecht (1998)
2. Horčík, R., Navara, M.: Validation sets in fuzzy logics. Kybernetika **38**(3), 319–326 (2002)
3. Pavelka, J.: On fuzzy logic I. Zeitsch. f. Math. Logik **25**, 45–52 (1979)
4. Pavelka, J.: On fuzzy logic II. Zeitsch. f. Math. Logik **25**, 119–134 (1979)
5. Pavelka, J.: On fuzzy logic III. Zeitsch. f. Math. Logik **25**, 447–464 (1979)
6. Goguen, J.A.: The logic of inexact concepts. Syntheses **19**, 325–373 (1968/69)
7. Turunen, E.: Well-defined fuzzy sentential logic. Math. Logic Q. **41**, 236–248 (1995)
8. Di Nola, A., Leustean, I.: Łukasiewicz Logic and MV-Algebras. In: Cintula, P., Noguera, C., Hájek, P. (eds.) Handbook of Mathematical Fuzzy Logic II. Studies in Logic, vol. 36, pp. 469–583 (2011)
9. Turunen, E., Navara, M.: Perfect Pavelka logic. Fuzzy Sets Syst. doi:10.1016/j.fss.2014.06.011
10. Suppes, P.: Introduction to Logic. Van Nostrand (1957)
11. Chang, C.C.: Algebraic analysis of many-valued logics. Trans. Am. Math. Soc. **88**, 467–490 (1958)

Index

A. Di Nola et al., *Fuzzy Logic of Quasi-Truth: An Algebraic Treatment*,
Studies in Fuzziness and Soft Computing 338, DOI 10.1007/978-3-319-30406-9

Printed in the United States
By Bookmasters